DIVERSIFYING MATHEMATICS TEACHING

ADVANCED EDUCATIONAL CONTENT AND METHODS
FOR PROSPECTIVE ELEMENTARY TEACHERS

DIVERSIFYING MATHEMATICS TEACHING

ADVANCED EDUCATIONAL CONTENT AND METHODS
FOR PROSPECTIVE ELEMENTARY TEACHERS

SERGEI ABRAMOVICH

State University of New York at Potsdam, USA

World Scientific

NEW JERSEY · LONDON · SINGAPORE · BEIJING · SHANGHAI · HONG KONG · TAIPEI · CHENNAI · TOKYO

Published by

World Scientific Publishing Co. Pte. Ltd.
5 Toh Tuck Link, Singapore 596224
USA office: 27 Warren Street, Suite 401-402, Hackensack, NJ 07601
UK office: 57 Shelton Street, Covent Garden, London WC2H 9HE

Library of Congress Cataloging-in-Publication Data
Names: Abramovich, Sergei.
Title: Diversifying mathematics teaching : advanced educational content and
    methods for prospective elementary teachers / by Sergei Abramovich
    (State University of New York at Potsdam, USA).
Description: New Jersey : World Scientific, 2017. | Includes bibliographical references and index.
Identifiers: LCCN 2017000441| ISBN 9789813206878 (hardcover : alk. paper) |
    ISBN 9789813208902 (pbk : alk. paper)
Subjects: LCSH: Mathematics teachers--Training of. | Elementary school teachers--Training of. |
    Mathematics--Study and teaching (Elementary)
Classification: LCC QA135.6 .A225 2017 | DDC 372.7--dc23
LC record available at https://lccn.loc.gov/2017000441

British Library Cataloguing-in-Publication Data
A catalogue record for this book is available from the British Library.

Cover design by Leonard Abramovich

Printed in Singapore

Preface

The book comprises advanced chapters of a mathematics content and methods course for prospective elementary teachers aimed at their familiarity with and understanding of ideas naturally emerging from the mathematics curriculum they prepare to teach. Most of the existing publications for prospective (and practicing) elementary teachers rarely demonstrate the unfolding intricacy of mathematical concepts beyond the traditional content. This inadvertent omission of a more advanced content of elementary mathematics leaves out many useful methods of teaching the subject matter. The main pedagogic idea of the book is to show that elementary mathematics content and its methods of teaching are not in the relation of dichotomy when methods do not depend on content and content does not affect methods. Rather, the book intends to show how the diversity of methods stems from the knowledge of content and how the appreciation of this diversity opens a window to the teaching of new (extended) content. Put another way, the book includes material that the author would have shared with teacher candidates had there have been more instructional time than a 3 credit hour master's level course "Elementary Mathematics: Content and Methods" provides. Thus, the book can be used to supplement a basic text for such a course by extending content and diversifying methods.

The ideas of the book have been developed by the author through teaching the above course for almost two decades to teacher candidates of the United States and Canada. The university where the author works is located in Upstate New York in close proximity to Ottawa and Montreal, and many of the author's students are Canadians pursuing their master's degrees in education. In addition to North American mathematics education standards, several other similar materials developed throughout the world and available in English are reviewed in the book as appropriate. Also, the book is informed by the author's participation in supervising pre-student teachers' work with young children in regular classrooms that often extends to various after school learning initiatives. Consequently, several articles published by the author as reflections on those initiatives inform content of some of the chapters.

Another aspect of the book concerns the demonstration of directions in which mathematics teacher education classroom discourse can develop, as teacher candidates are encouraged to ask questions by reflecting on the course content, each other work, and their own teaching engagements of different types. Likewise, the book demonstrates what kind of questions "mathematically proficient students" (the term used nowadays in the United States to embrace all students) may ask and how teacher candidates' own mathematical competence develops through learning to answer those questions. This shows true integration of content and methods of teaching elementary mathematics through which teaching methods not only are informed by a content but often expand conventional boundaries of the content.

The book consists of nine chapters. The first chapter reviews the modern day educational documents related to the teaching of primary school mathematics (including recommendations for mathematics teacher educators). These documents include Common Core State Standards (United States), The Mathematical Education of Teachers, I & II (United States), National Curriculum (England), National Curriculum in Mathematics (Australia), Primary Mathematics Teaching and Learning Syllabus (Singapore), The Ontario Curriculum (Canada), Elementary School Teaching Guide (Japan), National Mathematics Curriculum (Korea), Primary Mathematics Standards (Chile), and other documents. These materials are connected to contents of the chapters of the book and to several theories of teaching and learning mathematics. The main focus of the chapter is on answering and asking conceptual questions as the major means of teaching and learning mathematics as well as on the importance of skills connecting procedural and conceptual knowledge. The chapter also includes solicited reflections of teacher candidates on the international mathematics standards and teaching ideas they experience as they prepare to become teachers of young children.

The second chapter shows how the variation and growth of counting skills can be used to develop mathematical concepts discussed throughout the book. The main focus of the chapter is on the basic counting techniques of enumerative combinatorics. Most of those techniques stem from the rule of sum and the rule of product. The chapter emphasizes an approach to counting combinations (without and with repetitions) based on counting the number of permutation of letters in a word. The difference between counting techniques is discussed by comparing different real-life contexts. The importance of an experimental approach to the development of formulas using concrete materials is accentuated.

The third chapter demonstrates various uses of manipulative materials in the context of teaching elementary mathematics. It addresses several recommendations for mathematics teacher preparation and standards for teaching school mathematics. In particular, the chapter takes into account that "Mathematically proficient students ... bring two complementary abilities to bear on problems involving quantitative relationships: the ability to *decontextualize* – to abstract a given situation and represent it symbolically ... and the ability to *contextualize* ... [that is] to probe into the referents for the symbols involved" [Common Core State Standards, 2010, p. 6, italics in the original]. To this end, the chapter illuminates a didactic approach to teaching mathematics through modeling with manipulative materials by creating isomorphic relationships between informal and formal representations of mathematical situations. The approach is grounded into a pedagogical tradition that views children building formal knowledge on the foundations of their abilities that they use in an informal, intuitive manner. The chapter emphasizes the importance of an experimental component in the teaching of mathematics and explains the use of concrete materials through the lens of the theory of semiotic mediation.

Mathematics, with its origin in the study of number and shape, has evolved from concrete activities to abstract concepts by means of logical reasoning and mental computation. As known from history, the first mathematical problems stemmed from the contexts of counting using the principle of one-for-one correspondence. Later, the physical manipulation of objects and visual argumentation regarding the relationship among the objects led to the need for names describing specific properties of numbers. The fourth chapter shows that the teaching of mathematics can be organized as a transition from seeing and acting on concrete objects to describing the visual and the physical through culturally accepted mathematical notation. To this end, the chapter introduces the "we write what we see" (W^4S) principle as a pedagogic maxim to be used in the teaching of mathematics. The main focus of the chapter is on the multiplication and division of fractions using area model in the form of rectangular grids. The meaning of the "invert and multiply" rule as well as other procedures involving fractions are explained in terms of the change of unit.

The fifth chapter deals with several grade appropriate aspects of the representation of a positive integer as a sum of like numbers. In the modern mathematics classroom, using either physical or virtual manipulative materials, partitioning problems of that kind can be introduced even before the study of arithmetic, thereby providing young learners with many useful mathematical reasoning skills. The chapter shows how the proper use of the traditional teaching methods of elementary mathematics content can motivate the bottom to top approach to mathematics curricula through the study of various advanced concepts, including difference equations, mathematical proof, and combinatorial counting. The chapter also shows how one can use the concept of partition of integers into summands to demonstrate a possible learning trajectory spanning from visual and hands-on (experiential) activities to symbolic to computational ones and then back to symbolic and/or experiential activities but at a higher cognitive level.

The pedagogy of the sixth chapter incorporates the hidden mathematics curriculum framework which is based on the notion that many seemingly disconnected mathematical activities and problems scattered across the whole pre-college mathematics curriculum are connected through a common conceptual structure which is *hidden* from learners because of its intrinsic complexity. By the same token, many seemingly routine tasks, when explored beyond the boundaries of the traditional curriculum, can be used as windows to big ideas often obscured in the curriculum. In the context of mathematics teacher education, such extended explorations require a certain level of mathematical competence and intellectual courage on the part of the instructor. Furthermore, the complex nature of explorations requires the development of learning environments conducive to revealing hidden mathematical concepts to the teachers in a pedagogically appropriate format. Proceeding from a mundane task about finding the sums of consecutive positive integers (known as trapezoidal numbers), the chapter demonstrates how this framework can facilitate informed entries into the properties of the trapezoidal numbers for elementary teacher candidates. It is argued that teachers' familiarity with trapezoidal numbers and grade-appropriate properties of the numbers allows for many methods of teaching arithmetic to be used in a format which is appealing to young children.

The seventh chapter shows how commonly available digital tools can enhance the development of geometric ideas taught at the elementary level and advance the ideas of a computational experiment in the context of geometry. One of the topics directly linked to geometry is measurement, an activity that is situated at the very origin of mathematics. Because this activity is the simplest method of empirical analysis of geometric objects that brings about empirical evidence with a relative ease, it can be given strong emphasis through an exploratory study of geometry at the primary level. The chapter shows how one can use measurement facility of a dynamic geometry program such as *The Geometer's Sketchpad* as a way of elucidating geometric ideas. It demonstrates how empirical evidence can be utilized to generalize informal experience to formal concepts and then make use of the concepts for formal demonstration of properties of geometric objects. It is argued that this approach allows for the preservation of rigor as the main element of mathematical practice.

The eighth chapter reviews basic ideas associated with the probability strand, the notion of chance and its geometric interpretation, and the relationship between theoretical and experimental probabilities. It is shown how one can use a spreadsheet in calculating experimental probability of an event and then compare theory with experiment. The chapter uses a probabilistic perspective to revisit a number of problems from the previous chapters and to consider some classic problems associated with such games of chance as coin tossing and die rolling. Formal reasoning tools of probability theory are used to explain Bertrand's Paradox Box Problem and Monty Hall Dilemma. The chapter emphasizes the value of using history of mathematics in the teaching of probability theory.

The ninth chapter deals with counter-examples used as thinking devices in the teaching of mathematics at the primary level. While the literature on counter-examples is mostly devoted to the tertiary and secondary levels of mathematics curricula, these thinking devices can be used at the primary level as well. Three major goals of using counter-examples in mathematics education are highlighted: to help a learner conceptualize a certain mathematical property, to demonstrate that a certain mathematical statement lacks generality, and to avoid an uncritical application of a mastered problem-solving method as one enters a new conceptual domain. In this chapter, several examples supporting the above three goals and reflecting on the material of the preceding chapters is presented.

In conclusion, with much respect I wish to thank Ms. Rok Ting Tan for invaluable editorial guidance during my work on the book that, with deeply appreciated support and recommendations of anonymous reviewers, resulted in its publication. Also, I acknowledge being under obligation to express sincere gratitude to Prof. Yakov Yu. Nikitin for his careful reading of Chapter 8 and advising me on its content. Finally, I owe a special debt of recognition to my son Leonard Abramovich, a graduate student at Savannah College of Art and Design, for his meritorious professionalism in the design of the cover of the book.

Sergei Abramovich
Potsdam, NY

Contents

Preface v

Chapter 1 1
Teaching Elementary Mathematics: Standards, Recommendations
and Teacher Candidates' Perspectives 1
   1.1 Introduction 1
   1.2 Questions as the Major Means of Learning Mathematics 3
   1.3 Answering Questions Both Procedurally and
       Conceptually 6
   1.4 Connecting Algorithmic Skills and Conceptual
       Understanding 8
   1.5 Developing Deep Understanding of Mathematics
       Through Making Conceptual Connections 11
   1.6 Teaching and Learning to Think Mathematically 13

Chapter 2 17
Counting Techniques 17
   2.1 Introduction 17
   2.2 Rules of Sum and Product 19
   2.3 Tree Diagram and the Rule of Product 20
   2.4 Permutation of Letters in a Word 21
   2.5 Combinations without Repetitions 24
   2.6 Combinations with Repetitions 27

Chapter 3 33
Counting and Reasoning with Manipulative Materials 33
   3.1 Introduction 33
   3.2 Constructing a Triangle out of Straws 35
      3.2.1 Reflection on the activity with straws 38
      3.2.2 A real-life application of the triangle inequality 40
      3.2.3 Modifying the S^2AC^2 algorithm to enable
           linguistic coherency 41
      3.2.4 How many triangles can be constructed? 43

3.2.5   Using multicolored straws 45
        3.2.5.1   An equilateral triangle 46
        3.2.5.2   An isosceles triangle with the base being
                  the smaller side 47
        3.2.5.3   An isosceles triangle with the base being
                  the larger side 48
        3.2.5.4   A scalene triangle 49
3.3   Two Types of Representation as Means of Transition
       from Visual to Symbolic 49
3.4   Signature Pedagogy of Elementary Mathematics
       Teacher Education 51
3.5   Towards Rich Interpretations of Manipulative
       Representations 51
       3.5.1   Manipulative representation as text 51
       3.5.2   From "brothers" to Pascal's triangle 54
3.6   Learning to Move from One Type of Symbolism to
       Another and Back 56
3.7   Perimeter and Area Using Square Tiles 58
3.8   The Importance of Teacher Guidance in Using
       Manipulative Materials by Students 63
3.9   Conceptualizing Base-Ten System Using
       Manipulative Materials 65
3.10   Modeling as a Way of Creating Isomorphic
       Relationships 68

Chapter 4 71
We Write What We See (W⁴S) Principle 71
4.1   Introduction 71
4.2   W⁴S Principle and the Duality of Its Affordances 73
4.3   W⁴S Principle in Teaching Primary School Mathematics ... 74
4.4   Comparing Non-Unit Fractions Using Area Model 78
       4.4.1   Comparing fractions being close to each other 78
       4.4.2   Comparing fractions that are a unit fraction
               short of the whole 80
4.5   From Comparison of Fractions to Arithmetical Operations
       Using Area Model 81

4.5.1 Fractions as part-whole and divisor-dividend
models 81
4.5.2 The concept of common denominator 83
4.5.3 Reducing a fraction to the simplest form 84
4.5.4 Using unit fractions as benchmark fractions 84
4.5.5 Multiplying fractions using area model 87
4.6 Dividing Fractions Using Area Model 89
4.6.1 Partition model for division supports
contextualization 89
4.6.2 The importance of unit in solving word problems
with fractions 91
4.6.3 The meaning of "invert and multiply" rule 94
4.7 Ratio and Proportion 96
4.8 Percent and Decimal as Alternative Representations
of a Fraction 98
4.9 Multiplying and Dividing Decimal Fractions 100

Chapter 5 105
Partitioning Integers into Like Summands 105
5.1 Introduction 105
5.2 Partition of Integers into Summands 106
5.3 Activities with Towers Motivate Introduction of
Algebraic Notation 116
5.4 Ferrers-Young Diagrams 117
5.5 Recursive Definition of $P(n, m)$ Informed by
Ferrers-Young Diagrams 117
5.6 Making Mathematical Connections 120
5.7 Recursive Definition of $Q(n, m)$ Informed by
Ferrers-Young Diagrams 122
5.8 Connection to Triangular Numbers Opens a Window
to a New Concept 125

Chapter 6 129
Hidden Curriculum of Mathematics Teacher Education 129
6.1 Introduction 129
6.2 The Basic Task 130

6.3 Background Information: Triangular and Trapezoidal
      Numbers 132
      6.3.1 Activities 132
      6.3.2 Solutions to the tasks 133
      6.3.3 Trapezoidal numbers 137
6.4 Conceptually Oriented Discussion of the Basic
      Problem 139
      6.4.1 Grouping the sums by the number of addends 139
      6.4.2 Grouping the sums by the first addend 142
      6.4.3 Partitioning integers into the sums 142
6.5 Discourse Motivated by Multiple Ways of Creating
      Sums of Consecutive Natural Numbers 143
      6.5.1 Clarifying the meaning of the word special in the
             context of arithmetic 143
      6.5.2 The first encounter with a special property of the
             sums of two addends 144
      6.5.3 Moving from novice to expert practice in
             revealing special properties 146
      6.5.4 Exploring the sums of four consecutive integers 148
6.6 Proof of the Conjecture about Trapezoidal Numbers 151
6.7 Sums in Pairs of Odds and Evens 152
      6.7.1 Learning to generalize from special cases 153
      6.7.2 Comparing triangles to trapezoids with the top
             row greater than two 156
6.8 Mathematical Knowledge Used for Teaching
      Young Children 158
6.9 How Many Trapezoidal Representations Does
      a Number Have and How Can One Find Them? 159

Chapter 7 163
Informal Geometry 163
7.1 Introduction 163
7.2 Geoboard Explorations 165
7.3 Towards a Double-Application Environment 172
7.4 Guided Exploration on a Computational Geoboard 174
7.5 Transition to a Spreadsheet 178

7.6   Preparing Data for Empirical Induction 180

7.7   Abstracting from Numbers to Equations Using
       First-Order Symbols 182

7.8   Visual and Symbolic Deduction of Pick's Formula 183

7.9   Moving to a New Learning Site 187

7.10   Measuring vs. Counting 188

7.11   Encountering Limitation of the Environment 190

7.12   From Particular to General through Visualization 191

7.13   Communicating about Mathematics 194

Chapter 8 197

Probability as a Blend of Theory and Experiment 197

8.1   Introduction 197

8.2   Basic Concepts and Tools of the Probability Strand 200

       8.2.1   Randomness and sample space 200

       8.2.2   A more complicated example of constructing
               a sample space 202

       8.2.3   Different representations of a sample space 202

8.3   Fractions as Tools in Measuring Chances 205

8.4   Bernoulli Trials and the Law of Large Numbers 207

8.5   A Problem of Chevalier De Méré 209

8.6   A Modification of the Problem of De Méré 210

8.7   Wagering for the Odds/Evens in a Game of Chance 212

8.8   Paradoxes in the Theory of Probability 214

       8.8.1   Bertrand's Paradox Box problem 214

       8.8.2   Monty Hall Dilemma 215

8.9   Probabilistic Perspective on Partitioning Problems 217

       8.9.1   A problem of tossing three dice 217

       8.9.2   Unordered partitions of integers into unequal
               summands 218

8.10   Experimental Probability 219

       8.10.1 Experimental probability calls for a long
               series of observations 219

       8.10.2 Comparing experimental and theoretical
               probabilities when tossing a fair coin 220

8.10.3 Calculating relative frequencies for the problems
of De Méré 222
8.11 Exploring Irreducibility of Fractions through the
Lenses of Probability 224

Chapter 9 227
Using Counter-Examples in the Teaching of Elementary
Mathematics 227
9.1 Introduction 227
9.2 The Pedagogy of Using Counter-Examples 228
9.2.1 The role of linguistic constraints 228
9.2.2 A counter-example as a motivation for further
learning 231
9.3 Providing Explanation through Counter-Examples 231
9.4 Constructing a Counter-Example: an Illustration 233
9.4.1 From modeling with fractions to algebraic
generalization 233
9.4.2 From a counter-example to its conceptualization ... 234
9.4.3 A family of jumping fractions found
by a teacher candidate 235
9.4.4 Conceptualizing the teacher candidate's choice
of seven 237
9.5 A Counter-Example and Empirical Induction 238
9.6 Counter-Example as a Tool for Conceptual Development .. 240
9.7 Counter-Example in Explaining the Meaning of
Negative Transfer 241
9.7.1 Inequalities 242
9.7.2 Counting matchsticks 244
9.8 Transition from Combinations without Repetitions to
Combinations with Repetitions 246
9.9 Missing Fibonacci Numbers 248

Bibliography 251

Index 265

Chapter 1

Teaching Elementary Mathematics: Standards, Recommendations and Teacher Candidates' Perspectives

1.1 Introduction

This chapter aims to demonstrate the relevance of the content of the book to pedagogical ideas penetrating standards for teaching mathematics and recommendations for the mathematical preparation of teachers that have been the backbone of the current reform movement of elementary mathematics teaching worldwide. The author's current professional experience includes teaching mathematics to elementary teacher candidates of the United States and Canada[1] as well as involvement in various service-related mathematics education endeavors internationally. Both the teaching and the service created conditions for the study and appreciation of mathematical standards used in North America and elsewhere in the world. As will be demonstrated through multiple citations of the standards (available in English) for mathematics teaching at the elementary level and expectations for the teachers of young students, there is a true congruency of the ideas of mathematics pedagogy be it Australia, Canada, Chile, England, Japan, Korea, Singapore, or the United States. Throughout this and other chapters of the book, the author will share solicited comments by elementary teacher candidates about their own learning to teach mathematics at the primary level that indicate the candidates' interest in acquiring deep understanding of grade-appropriate mathematical content and methods towards gaining much needed professional expertise. Such interest stems from their passion to help schoolchildren to be better mathematical

---

[1] As mentioned in the Preface, the university where the author works is located in the United States in close proximity to Canada, and many of the author's students are Canadians pursuing their master's degrees in education. This diversity suggests the importance of aligning teacher education courses, in general, and mathematics education courses, in particular, with multiple international perspectives on the teaching of mathematics.

thinkers because, as one of them put it, *"If a student makes an error or misunderstands a mathematical concept or procedure, a teacher with deep understanding of the material will be able to recognize why the error was made and correct the student's thinking without imposing a step-by-step procedure on how to solve the problem"*. In other words, it is at least equally important to know *how* to solve a problem and *why* the problem-solving procedure is the correct solution.

There is also a consensus among mathematics educators worldwide that "teachers must know the mathematical content they are responsible for teaching not only from a more advanced perspective but beyond the level they are assigned to teach" [Baumert *et al.*, 2010, p. 137]. This, in particular, applies to elementary teacher candidates who, thereby, should be given multiple opportunities to acquire strong knowledge and 'deep understanding' of mathematics through the appropriate mathematical and pedagogical preparation within mathematics content and methods courses that are mandatory for any teacher education program. The total number of relevant mathematics courses may differ from program to program and may include elective content courses to satisfy mathematics prerequisites of the candidates. Internationally, researchers found that "programs that compromise on subject matter training ... have detrimental effects on PCK [pedagogical content knowledge] and consequently negative effects on instructional quality and student progress" [*ibid*, p. 167]. One should not underestimate the importance of subject matter training because, as Chilean mathematics educators put it, "many aspiring teachers have major gaps in [mathematical] knowledge and skills when entering [elementary] teacher education programs" [Felmer *et al.*, 2014, p. 158].

At a local level, one of the author's students, when asked to reflect on what the researches found, shared a belief developed through spending a semester as an intern-observer at an elementary school: *"teachers who receive formal training on subject matter content are able to implement strategies and teaching methods in a more effective way that increases student outcomes"*. Another teacher candidate suggested, *"teachers should be specifically trained for math as well as other subjects on what content they should be teaching for each grade level"*. Indeed, as yet another teacher candidate questioned, *"if teaching programs compromise on subject matter training then where would that knowledge come from?"* In that way, the above-mentioned research finding and the cited comments by teacher candidates provided motivation for writing this book the goal of which is to show how the traditional curriculum of an elementary mathematics content and methods course can be extended to lend more weight to subject matter training.

1.2 Questions as the Major Means of Learning Mathematics

What does it mean that elementary teacher candidates need to possess 'deep understanding' of mathematics? Why do they need to have such understanding? There are several reasons for prospective teachers to be thoroughly mathematically prepared in order to have positive effects on the progress of young learners of mathematics. First, in the modern mathematics classroom, students of all ages are expected and even encouraged to ask questions. In the United States, the national standards already for grades pre-K – 2 suggest, "Students' natural inclination to ask questions must be nurtured ... [even] when the answers are not immediately obvious" [National Council of Teachers of Mathematics, 2000, p. 109]. Support to this suggestion can be found in the following comment by a teacher candidate: *"It is okay not knowing the answer to the question but it is not okay with leaving that question unanswered"*. The candidate describes herself as *"the type of educator that will always encourage my students to ask themselves some of those same questions that will allow them to participate in some profound thinking"*.

Just across the border with the United States, the Ontario Ministry of Education in Canada, through their mathematics curriculum for early grades, sets expectations for teachers to be able to "ask students open-ended questions ... encourage students to ask themselves similar kinds of questions ... [and] model ways in which various kinds of questions can be answered" [Ontario Ministry of Education, 2005, p. 17]. In order to develop such proficiency, "teachers should know ways to use mathematical drawings, diagrams, manipulative materials, and other tools to illuminate, discuss, and explain mathematical ideas and procedures" [Conference Board of the Mathematical Sciences[2], 2012, p. 33]. In Chile, mathematics teachers are expected to "use representations, call on prior knowledge, put forward good questions, and stimulate an inquisitive attitude and reasoning among students" [Felmer *et al.*, 2014, p. 37]. In Australia, mathematics teachers know how to motivate "curiosity, challenge students' thinking, negotiate mathematical meaning and model mathematical thinking and reasoning" [Australian Association of Mathematics Teachers, 2006, p. 4]. The repertoire of learning opportunities the teachers offer to their students includes continuous search for alternative approaches to solving problems as well as helping students to better learn a specific problem solving strategy with which they have been struggling. National mathematics curriculum in England uses such terms as "practice with increasingly complex problems over time … [and] can solve problems … with increasing sophistication" [Department for Education, 2013, p. 1]. Towards this end, teachers have to be prepared to deal with situations when natural quest for inquiry leads students towards this sophistication and increase in complexity of mathematical ideas. The need for this kind of teacher preparation is confirmed by a teacher candidate who put it as follows: "*If a student asks why, and a teacher cannot explain how something has come to be, the student loses all faith and interest in the subject and respect for the teacher*".

The chapters of this book are aimed at preparing elementary teacher candidates to have the required experience through their own study of mathematics within an elementary teacher preparation program that is sometimes limited to a single methods and content course in mathematics. Even within a single course, attending which prompted another teacher candidate to remark, "*I have had many "lightbulbs" during class in which things have been made clearer to me, such as why three times three equals nine*", teacher candidates can be shown how the evolution of ideas, from simple to complex, can be a part of a classroom discourse related to basic operations of arithmetic. Perhaps, the "lightbulbs" comment was due to a classroom discussion that three times

---

[2]Conference Board of the Mathematical Sciences is an umbrella organization comprised of 16 professional societies in the United States concerned (in particular) with the mathematical preparation of schoolteachers.

three is not only a product of two equal numbers, but also the sum of three equal numbers, the sum of the first three odd numbers, the sum of three consecutive natural numbers (Chapter 6, section 6.3.3), and, to continue emphasizing the three, the sum of consecutive *triangular* numbers (Chapter 6, section 6.3.2).

Likewise, it is not difficult to partition a natural number (often colloquially called a counting number) in two differently ordered like summands and furthermore, in the case of a large number, to find the number of such partitions without making an organized list. Towards this end, one can check to see that a natural number n can be partitioned in two differently ordered natural numbers in $n - 1$ ways. For example, $5 = 1 + 4, 5 = 4 + 1, 5 = 2 + 3, 5 = 3 + 2$ – four partitions; and $4 = 1 + 3, 4 = 3 + 1, 4 = 2 + 2$ – three partitions. One can see that whereas the (odd) number 5 cannot be partitioned in equal integer summands, the (even) number 4 does have such partition. Therefore, in general, two cases need to be considered. When $n = 2k$ (i.e., n is an even number) we have $2k - 1$ partitions (with regard to the order of summands):
$$1 + (2k - 1) = (2k - 1) + 1 = \mathbf{2} + (2k - 2) = (2k - 2) + 2 = ... = (\mathbf{k - 1}) + (k - 1) = (k + 1) + (k - 1) = \mathbf{k} + k.$$
When $n = 2k - 1$ (i.e., n is an odd number) we have $2k - 2$ partitions (with regard to the order of summands):
$$1 + (2k - 2) = (2k - 2) + 1 = \mathbf{2} + (2k - 3) = (2k - 3) + 2 = ... = (\mathbf{k - 1}) + k = k + (k - 1).$$
By not counting sums with the same summands twice, we have k unordered partitions in the former case and $k - 1$ unordered partitions in the latter case (the counters of such partitions are in the bold font); in other words, when n is even, the number of unordered partitions is $n/2$ and when n is odd, the number of unordered partitions is the integral part of $n/2$.

But already finding the number of partitions of a relatively small natural number into three like summands presents a problem with no immediately available answer[3]. Yet, it is a characteristic feature of mathematics that frequently, a simply formulated question or its natural extension might not have an easy (or, at least, an immediate) answer. It is through seeking answers to such questions that mathematical knowledge

---

[3] An answer to this question in the case of unordered partitions is given in Chapter 5.

develops. A contextualization of the last question is an easy effort: for example, finding all ways to put 10 apples in three baskets (not leaving any basket empty). What about four baskets? Questions of that kind are easy to understand and not easy to answer. It is a skill of asking an easy to understand question that motivates queries the mathematical complexity of which may not be realized by a student and even by a teacher. Teachers, however, have to be prepared to help students realize the mathematical complexity of a question through the process of de-contextualization showing that what mathematics is all about is to satisfy natural curiosity of the (ever inquiring) mind. The importance of such realization is two-fold: it elevates students' self-confidence as owners of 'good questions' and helps connecting context with content. As a teacher candidate put it, "*I want my students to feel comfortable taking risks and asking questions because they need to be able to relate to the mathematics material and the best way for them to do that is to take some responsibility for their own education*".

Also, a search for an alternative procedure in solving a mathematical problem quite unexpectedly may become a source of a new conceptual development and insight. For example, as will be shown in Chapter 6, after decomposing integers into the sums of consecutive natural numbers based on the number of addends in the sum, one can be asked about an alternative procedure in order to diversify the methods of teaching. In doing so, one can group the decompositions by the number of addends. As a result, one can come to recognize, among other things, that a sum with an odd number of addends is a multiple of that number. Likewise, one can group decompositions by the first addend. In that case, one can realize that not all integers can be represented as the sums of consecutive natural numbers. In addition, by utilizing the diversity of thinking among students as a means of individual intellectual growth, one has an independent confirmation that all the sums have been found as "it is independency of new experiments that enhances credibility" [Freudenthal, 1978, p. 193].

1.3 Answering Questions Both Procedurally and Conceptually

Often, an additive or a multiplicative decomposition of natural numbers in two like summands or two like factors, respectively, as a way of learning mathematical facts from addition and multiplication tables constitute the main mathematical experience of elementary teacher candidates as they enter a teacher preparation program. Yet, already in the early grades, the goal of mathematics instruction "is to help children attain both computational proficiency and conceptual understanding" [Conference Board of the Mathematical Sciences, 2001, p. 15]. Why don't we have subtraction or division tables? Is it because of the need to reduce the number of facts for students to memorize? Or, is there a conceptual basis for the absence of such tables in the curriculum? Having experience in asking and answering conceptually-oriented questions associated with the natural extension of seemingly mundane procedures is useful not only because of their relevance to the entire primary school mathematics curriculum but such experience is necessary for acquiring a rich toolkit of content knowledge and pedagogical skills "emphasizing the importance of a systematic path of mathematics" [Takahashi *et al.*, 2004, p. 11], as mathematics educators in Japan believe. By learning to ask 'good questions' or answering such questions, prospective teachers "can scaffold learning ... correct a misconception, reinforce a point or expand on an idea" [Ministry of Education Singapore, 2012, p. 27].

For example, in the context of a geoboard (Chapter 7), through an additive decomposition of the number 10 in two natural numbers one can discover that the equality $10 = 1 + 9$ brings about a rectangle of perimeter 20 linear units with the smallest area. Likewise, through a multiplicative decomposition of the number 20 in two natural numbers one can discover that the equality $16 = 1 \times 16$ brings about a rectangle of area 16 square units with the largest perimeter. But outside a geoboard, which allows for decompositions into non-integer summands/factors, does there exist a rectangle with a given perimeter/area and the smallest/largest area/perimeter? Expanding on the idea of decomposition and the concept of area, one can correct a possible misconception about the existence of such rectangles. In other words, "new concepts come through the shift of old concepts to new situations" [Schon, 1963, p. 53]. Proceeding from multiple ways of additive/multiplicative decompositions of a natural number in two like summands/factors, one such new situation can be naturally brought to bear by students in the elementary classroom who, in the quest "to uncover abstract mathematical concepts or results ... investigate whether rectangles with the same perimeter can have different areas" [Ministry of Education Singapore, 2012, p. 23]. Consequently, because "nothing happens in this world in which some reason of maximum or minimum would not come to light" (Euler[4], cited in [Pólya, 1954]), a question about a rectangle with the smallest area or the largest perimeter can be raised and a teacher has to be prepared to address it in a grade-appropriate, yet mathematically accurate way. This is an example of how "in the attempt to subsume the new situation [i.e., considering rectangles the side lengths of which become smaller and smaller] under the old concept [i.e., area as the number of unit squares] may be the basis of the new concept" [Schon, 1963, p. 27], something that justifies the need for conceptualizing the area of a rectangle as the product of its length and width. By having mathematical questions about naturally unfolding situations answered both procedurally and conceptually, students would appreciate "opportunities to extend their learning ... [through] tasks that stretch their thinking and deepen their understanding" [Ministry of Education Singapore, 2012, p. 25].

1.4 Connecting Algorithmic Skills and Conceptual Understanding

The second reason for elementary teacher candidates having strong mathematical preparation is due to an incorrect assumption, as noted by the Conference Board of the Mathematical Sciences [2012], "that prospective elementary teachers ... learn all the mathematics they need to teach mathematics well during their own schooling" (p. 5). The following comment by a teacher candidate explains the fallacious nature of a belief in the simplicity of teaching elementary mathematics: *"When I came into this class I fully understood the concept of subtraction, however, I was not fully equipped with the skills one would need to teach this concept to a first and second grade classroom. Additionally, I had not put together that there were different types of subtraction problems that I could pose to the class. Because of this, I was not prepared to go into a classroom and actually teach children how to subtract. With the strategies I have in place now, I believe that I am much more prepared to*

---

[4] Leonhard Euler (1707–1783) – a Swiss mathematician, the father of all modern mathematics.

teach mathematical concepts than I was before taking this class". In the modern day mathematics methods and content courses, elementary teacher candidates need to acquire what is called mathematical knowledge for teaching (e.g., [Adler and Venkat, 2014]) or, more generally, pedagogical content knowledge – a notable educational construct originally introduced by Shulman [1986]. It is through continuously reflecting on the material learned, by being engaged in asking and answering a variety of questions by using different types of manipulative materials (both physical and virtual) as thinking devices that, in the context of mathematics education, such knowledge gradually develops. Whereas "learning to teach is a matter of learning to explain procedures clearly and assembling a toolkit of tasks and activities to use with children" [Conference Board of the Mathematical Sciences, 2012, p. 34], even such a simple procedure as long division, in order to be clearly explained on the conceptual level, needs to be connected to common sense.

For instance, when dividing 3 into 741 using long division, one starts with dividing 3 into 7 (the first digit) to get the quotient two and the unit remainder which then joins 4 (the second digit) to become 14; the latter number when divided by 3 gives the quotient 4 and the remainder 2 which then joins 1 (the last digit) to become 21 – a number evenly divided by 3 to give the quotient 7. This formal rule of obtaining 247 as a result of long division reflects common sense through contextualizing the procedure as distributing evenly, say, 741 apples packed in boxes of 100 and 10 apples among three stores. Without using any formal mathematics, one first deals with seven large boxes (assigning two boxes, not opening them, to each of the three stores) and then unpacking the remaining box within which apples are packed in 10-apple boxes and dealing with 14 of them (assigning 4 boxes to each of the three stores), and then finally unpacking the remaining two smaller boxes to distribute 21 apples among three stores. Nowhere in this physical process an arithmetical concept called by mathematics educators the partition (or partitive) model for division was used, yet this model stems from the de-contextualization of action which, in turn, stems from common sense. This shows the importance of preparing elementary teacher candidates not only to hold procedural mastery needed to obtain a certain information (e.g., in the context of long division) but to be able to explain specific aspects of their mastery that enable computational procedures to work (e.g., to explain the order in which the divisor is applied to the face values of a multi-digit dividend). Reflecting on interplay between the procedural and the conceptual, a teacher candidate explained that understanding of how one feeds into the other would help her not to teach through the stance *"it is just because I said that is how you do it. It is really giving the children the chance to understand why we do certain things. When you teach math like that it becomes less scary"*.

Therefore, operations have to be thoroughly motivated and explained both procedurally and conceptually. Indeed, teachers of mathematics are expected to teach by "offering students opportunities to learn important mathematical concepts and procedures with understanding" [National Council of Teachers of Mathematics, 2000, p. 3]. In Chile, mathematics teachers appreciate the notion that "concepts are inseparable from their representations ... [the variety of which] must be provided for a particular concept, taking into account the relationships among representations ... [something that] helps students achieve a deeper understanding of the concept and provides more tools for working with it" [Felmer *et al.*, 2014, p. 35]. Likewise, as noted by Hwang and Han [2013], mathematics teachers in Korea are expected to "emphasize not only the results of problem solving but also its strategies and processes and encourage students to formulate problems on their own" (p. 36). Yet such teaching skills are not on the surface. In order to develop skills that unite procedural and conceptual knowledge, teacher candidates need to possess conceptual understanding of a problem regardless it can be solved by applying a well-defined algorithm. This kind of understanding is critical for enabling a mathematics classroom to be a place where "children can learn, without drill, to deal empirically with situations involving numbers and develop a flexible set of procedures for handling such routine as is necessary" [Association of Teachers of Mathematics, 1967, p. 6].

As an example of extracting a concept from a procedure, once again, consider the algorithm of division. Whereas, in general, using an algorithm does not necessarily lead to its conceptual understanding, a special case may motivate making a connection between well-developed procedural skills and emerging conceptual knowledge. To this end, one can be asked to construct a sequence the first two terms of which are equal to one and each other term is the sum of the previous two terms. In that way, the sequence 1, 1, 2, 3, 5, 8, 13, 21, 34, 55, ... can be constructed. In the sequel, one can be asked to develop another sequence comprised of the remainders generated through dividing two consecutive terms of the constructed sequence, the larger by the smaller, beginning with the pair (2, 3). Through this process, once again, the sequence 1, 2, 3, 5, 8, 13, 21, ... emerges. Then, one can be asked to explain why the division produced the same sequence (starting from the second term). By answering this question, one can realize that because of the rule through which the former sequence has been developed, dividing the larger number by the smaller number always yields the unit quotient. This, in turn, brings about conceptual understanding of the resulting relationship among the quantities used in the process of division: the dividend equals the divisor times the quotient plus the remainder. Therefore, one way to connect two types of knowledge, procedural and conceptual, could be through the recourse to a special case. If necessary, the concepts of Fibonacci numbers and Euclidean algorithm (finding the greatest common divisor of two integers) can be introduced as an expansion of the idea of the conceptual meaning of division – the most complicated arithmetical operation among those studied at the elementary level. Therefore, as a teacher candidate opined when reflecting on this kind of a learning environment, *"Not only do we have to have a deep level of knowledge and understanding, but we also need to be able to ask deep and open-ended questions. This will increase the students' level of problem solving abilities, as well as their creative and mathematical thinking"*.

1.5 Developing Deep Understanding of Mathematics Through Making Conceptual Connections

The instructional shift that Common Core State Standards [2010], the major educational document in the United States at the time of writing this book, called coherence, requires that teachers "know how the mathematics they teach is connected with that of prior and later grades"

[Conference Board of the Mathematical Sciences, 2012, p. 1]. The notion of connections in mathematics education has many facets and the rich interplay that exists among the concepts of mathematics is one such facet. Therefore, the third reason for providing elementary teacher candidates with mathematical knowledge beyond basic operational proficiency is to develop the appreciation of the notion of mathematical (conceptual) connections that is important because it is a cornerstone of "the ability to justify, in a way appropriate to the student's mathematical maturity, *why* a particular mathematical statement is true or where a mathematical rule comes from" [Common Core State Standards, 2010, p. 4, italics in the original]. As educators in England suggest, students "must be assisted in making their thinking clear to themselves as well as others, and teachers should ensure that pupils build secure foundations by using discussion to probe and remedy their misconceptions" [Department for Education, 2013, p. 2]. To this end, "mathematics courses and professional development for elementary teachers should not only aim to remedy weakness in mathematical knowledge, but also help teachers develop a deeper and more comprehensive view of the mathematics they will or already do teach" [Conference Board of the Mathematical Sciences, 2012, p. 23].

As will be emphasized in various chapters of this book, the appreciation of mutual connections among mathematical ideas can be developed by demonstrating to prospective elementary teachers through grade appropriate activities that one of the common threads permeating the entire school mathematics curriculum (and consequently, the teacher preparation mathematical course work) is the representation of numbers as sums of other numbers. Similarly to what was already shown above, integers (not all though) may be represented through the sums of consecutive natural numbers; squares of integers – through the sums of odd, triangular, and square numbers; unit fractions – through the sums of like fractions; integers – through irrational numbers; real numbers – through complex numbers; and so on. What many teacher candidates (in the author's experience) do not appreciate (or simply not aware of) is the significance of the fact that such representations may not be unique or do not exist at all. Such a diversity in representation of numbers calls for the diversity of methods of teaching the subject matter. Consequently, mathematics methods and content courses should include the variety of teaching strategies that reflect the broad mathematical content of elementary mathematics. Teachers need to learn how "to provide a more engaging, student-centered, and technology-enabled learning environment, and to promote greater diversity and creativity in learning" [Ministry of Education Singapore, 2012, p. 17]. Of a special noteworthiness is the stance that "teacher knowledge should be particularly important for the learning gains of weaker students" [Baumert *et al.*, 2010, p. 147]. Indeed, it is through the knowledge of mathematics that the diversity of teaching methods, to the benefit of struggling students, unfolds. Thinking about the future, a teacher candidate confessed, "... *it will be difficult at first to keep my mouth shut when I want to help a struggling student, but I need to keep in mind that it is that struggle which builds better math minds. My struggling students will be encouraged to solve mathematical problems using physical manipulatives, pictures, and by working collaboratively with other students*".

1.6 Teaching and Learning to Think Mathematically

It has been already half a century since it was commonly understood around the world and, in particular, widely accepted by the teachers of mathematics in England that "teaching which tries to simplify learning by emphasizing the mastery of small isolated steps does not help children, but put barriers in their way" [Association of Teachers of Mathematics, 1967, p. 3]. This decrease in focus on the memorization of procedures can be found in the modern day elementary mathematics curriculum in Japan explaining that when "students are forced to memorize steps of calculation mechanically, without understanding the meaning, or if teaching is done in a way that only emphasizes the formal processing of calculations, the value of knowledge and skills are significantly reduced" [Takahashi *et al.*, 2004, p. 39]. Even such a simple procedure as counting requires conceptual understanding. For example, an automatic counting of points on a circle, in the absence of acumen that the first point counted has to be singled out in order to have one-to-one correspondence between the points counted and the number names used, may lead to either a confusion or to an erroneous result. Such a naive counting of points on a circle may be used as a counter-example of the universal acceptance of the primordial character of procedural skills without a rather obvious codicil that no procedural skill is possible without having some basic conceptual understanding of the skill.

In order for students to develop basic mathematical skills, such skills "should be taught with an understanding of the underlying mathematical principles and not merely as procedures" [Ministry of Education Singapore, 2012, p. 15]. For example, if a student needs to find the product 5×7 and doesn't remember the corresponding multiplication fact, conceptual understanding of the operation as repeated addition and knowing the meaning of its factors, namely, that the first factor shows the number of repetitions of the second factor yields $5 \times 7 = 7 + 7 + 7 + 7 + 7$. Yet, counting by sevens is kind of difficult, unlike counting by fives, something that is taught already at the preoperational level. This insight allows one to take another step towards using conceptual understanding in simplifying procedural performance: utilizing commutative property of multiplication, the repeated addition of sevens can be replaced by the repeated addition of fives, $5 \times 7 = 7 \times 5 = 5 + 5 + 5 + 5 + 5 + 5 + 5$. Here one can see how emphasis on conceptual understanding of arithmetical operations, that is, using the commutativity of multiplication as a problem-solving tool, makes it "possible to change students' views from a fixed mind-set [being fixed on counting by sevens] to a growth mind-set [instead, counting by fives] in ways that encourage them to persevere in learning mathematics and improve achievement test scores as well as grades" [Conference Board of the Mathematical Sciences, 2012, p. 9]. The notion of a fixed mind-set vs. a growth mind-set can be understood not only in terms of changeable beliefs about one's abilities to do mathematics and the levels of achievement in completing tests but, not less important, in terms of automatism vs. insight in mathematical problem solving as "sources of insight can be clogged by automatisms" [Freudenthal, 1983, p. 469]. The automatism of seeing the first factor, 5, as a number of repetitions of the second factor, 7, perhaps afforded by a rigorous instruction, might prevent one from using insightful strategy of swapping the factors that enables a true mathematical elegance of counting by fives.

Teaching for understanding should help "students to have a clear purpose ... to understand the meaning of quantities and geometric figures, to enhance their ability to think, to make decisions, ... to feel the joy and meaning of learning mathematics" [Takahashi *et al.*, 2004, p. 17]. For example, the meaning of every number located on the main (top left – bottom right) diagonal of the multiplication table is, rendered simply, the product of two equal integer factors. Furthermore, a transition from one such number (product) to the next number requires augmentation by an odd number, which is one greater than twice the factor. This observation can be explained in geometric terms – a transition from the $n \times n$ square to the $(n+1) \times (n+1)$ square requires the extension of a pair of adjacent sides by the $1 \times n$ rectangles and then connecting these rectangles with the unit square. In other words, this transition can be expressed through one the most basic algebraic identities, $n^2 + 2n + 1 = (n+1)^2$. Making connections between relations among numbers and their geometric and algebraic interpretations brings about meanings to and understanding of the three entities of mathematics – arithmetic, geometry, and algebra. As a teacher candidate noted, *"The conceptual level of math is what grows from the deeper meaning. It is important that our children are being taught on a conceptual level and for this to happen, our teachers must comprehend the lessons on a deeper level"*. Toward this end, manipulative materials (e.g., square tiles) can be used to demonstrate such integrated perspective on the numbers in the multiplication table, the basic geometric figures, and algebra as a generalized arithmetic. That is, using concrete materials to investigate patterns in the multiplication table at a lower level can be revisited at a higher level in the context of teaching and learning geometrically enhanced algebra.

In order to be able to provide students with help in making connections through integrating different curricular strands, "teachers need to have the big picture in mind so that they can better understand what ... to do at their level, as well as to plan and advise students in their learning of mathematics" [Ministry of Education, Singapore, 2012, p. 11]. Such instruction should be engaging so that students, as mathematics educators in England suggest, can develop an appreciation "that there was more than one way of doing things" [Advisory Committee on

Mathematics Education, 2007, p. 18] as they reflect on their problem-solving experience. In particular, the use of concrete materials provides diverse opportunities for teaching and learning mathematical concepts and procedures with understanding. As a result, as Canadian mathematics educators believe, "the strategies teachers employ will vary according to both the object of the learning and the needs of the students" [Ontario Ministry of Education, 2005, p. 24]. Likewise, when encountering diverse learners of mathematics, teachers in Korea are expected to possess skills to "make questions as open as possible to allow students to solve a problem in a variety of ways ... as part of the overall effort to stimulate thinking on the part of the students" [Hwang and Han, 2013, p. 36]. It is through learning to think mathematically that elementary teacher candidates develop skills of teaching mathematics for understanding. The chapters that follow include material encouraging productive mathematical thinking, appropriate procedural mastery, and deep conceptual understanding.

Chapter 2

Counting Techniques

2.1 Introduction

Counting is one of the most basic mathematical skills. Being procedural in nature, it gives birth to many concepts that form the foundation of mathematics. For example, knowing that the order in which objects are counted or their different arrangements do not change the total count leads to the following big idea of mathematics – numbers can be represented as sums of other numbers in many different ways. Consider the set of square tiles shown in Fig. 2.1. When asked to find the number of tiles, one can count them one by one to give the answer nine. However, one can count tiles by rows (from top to bottom) not only to have the relation $1 + 3 + 5 = 9$, but to recognize that the sum in the left-hand side of this relation comprises the first three odd numbers and that its right-hand side is the square of three or, alternatively, three repeated three times. Consequently, the tiles in Fig. 2.1 can be arranged to form a square array shown in Fig. 2.2. The next counting engagement with the tiles could be to find all ways to put them in two, three, four, five, six, seven, eight, and nine groups. As will be shown in Chapter 5, the total number of ways to represent the number 9 as a sum of positive integers without regard to their order is 30. In a mean time, one can check to see that there are: one way – as a sum of nine integers, one way – as a sum of eight integers, two ways – as a sum of seven integers, three ways – as a sum of six integers, five ways – as a sum of five integers, six ways – as a sum of four integers, seven ways – as a sum three integers, four ways – as a sum of two integers, and one self representation. For example, in the case of five integers, we have

$$9 = 1+1+1+1+5 = 1+1+1+2+4 = 1+1+1+3+3$$
$$= 1+1+2+2+3 = 1+2+2+2+2 \, .$$

Another counting engagement in mathematics is to find the number of different arrangements of objects within a given set. For example, as all the nine tiles have different patterns (Figs. 2.1 and 2.2), one can count the number of their different patterned arrangements. One can also count the number of ways two tiles can be selected out of nine tiles, distinguishing between the tiles being unique objects and tiles being just types (allowing for the selection of the same type twice). The last two problems belong to the branch of mathematics known as combinatorics. This chapter deals with the development of some basic counting techniques of combinatorics using various real-life contexts. Classroom observations indicate that knowing such techniques is useful for the teachers of elementary school mathematics as young children frequently ask counting questions that cannot be answered experimentally, unlike finding all unordered partitions of nine in five summands. Developing different counting techniques required by natural extensions of the elementary mathematics content reveals the diversity of methods used in the process of conceptualization of the techniques and specific operations they incorporate.

Combinatorics is one of the oldest branches of modern mathematics and it goes back to the 16th century when the games of chance played an important role in the life of society [Vilenkin, 1971]. The need for the theory of such games stimulated the creation of specific counting techniques and mathematical concepts related to the new real-life situations. Further scholarly efforts of Pascal[5] and Fermat[6] in the pursuit of theoretical studies of combinatorial problems laid a foundation for the theory of probability (Chapter 8) and provided approaches to the development of combinatorics as the study of methods of counting various combinations of elements of a finite set.

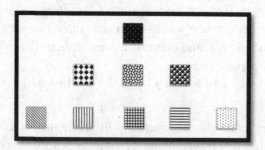

Fig. 2.1. Nine as 1 + 3 + 5.

---

[5] Blaise Pascal (1623–1662) – a French mathematician, physicist, and philosopher.

[6] Pierre de Fermat (1601–1665) – a French mathematician and lawyer.

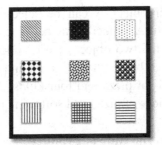

Fig. 2.2. Nine as 3 + 3 + 3.

As mentioned in a memorandum [Ahlfors, 1962] signed by 75 North American mathematicians – an early concern of professional mathematicians about didactics of mathematics – instruction should be a process helping one "to recognize a mathematical concept in, or to extract it from, a given concrete situation" (p. 190). Combinatorics is an appropriate context for accommodating such instructional philosophy. What kind of combinatorial reasoning is involved when one counts the number of ways books can be arranged on a shelf or selected from a shelf? How does one distinguish among counting situations related to the library, the grocery store, or the touchtone phone? While these are real-life situations that teacher candidates are familiar with, they need help in making sense of mathematics that is behind each situation so that they can extract the right concept from the appropriate context. Using a toolkit of mathematical teaching methods among which is recourse to a suitable contextualization can provide such help.

2.2 Rules of Sum and Product
There are two basic rules that govern counting in the context of combinatorics: the rule of sum and the rule of product.

The rule of sum: If object A can be selected in m ways and, independently from this selection, object B can be selected in n ways, then there are $m + n$ ways to select either A or B. For example, if Jane was accepted to three private and four state schools to become a teacher, then there are $3 + 4 = 7$ possible school choices that she can make.

The rule of product: If object A can be selected in m ways and if, following the selection of A, object B can be selected in n ways, then the *ordered* pair (A, B) of the two objects can be selected in mn ways. For example, if Jake wants to buy a slice of pizza and a soft drink for lunch and there are three kinds of pizza and four kinds of drink, then there are $3 \times 4 = 12$ possible choices of pizza and soft drink that he can make.

2.3 Tree Diagram and the Rule of Product

A tree diagram is a multiplicative structure used to organize counting according to the rule of product. Consider a problem of counting the number of different three-cube towers constructed out of three colors without repetition of colors. The cube at the bottom of such a tower can be selected in three ways; following its selection, the cube in the middle can be selected in two ways (thus the first two cubes can be selected in $3 \times 2 = 6$ ways), and the top cube, following the selection of the first two, can be selected in one way only. So, three-cube towers can be selected in $3 \times 2 \times 1 = 6$ ways.

Due to the commutative property of multiplication[7] we have $3 \times 2 \times 1 = 1 \times 2 \times 3$. The product $1 \times 2 \times 3$ has a special notation, $3!$ (reads "three factorial"), that is, $1 \times 2 \times 3 = 3!$ In general, the notation $n!$ (reads "n factorial") is used to represent the product of the first n counting numbers: $1 \times 2 \times 3 \times ... \times n = n!$.

However, if colors may be repeated, each of the three cubes can be selected in three ways and, therefore, there exist $3 \times 3 \times 3 = 27$ ways to construct a three-cube tower out of cubes in three colors. The use of a tree diagram depends on a context. For example, if one assigns a numeric value to a color, say 1 to red (R), 5 to green (G) and 10 to yellow (Y), then each of the six towers – RGY, RYG, GRY, GYR, YRG, and YGR – has the same numerical value, 16 (= 1 + 5 + 10). Consequently, to find through a tree diagram the total number of sums of money made out of three coins – pennies, nickels, and dimes – is an incorrect strategy. As will be shown below, out of 27 towers that can be constructed out of

---

[7] Commutative property of multiplication and commutative property of addition have conceptually different methods of demonstration: while the latter is based on the irrelevance of the order in which objects are counted, the latter requires recourse to geometry and is based on the invariance of area under rotation.

three cubes in three colors only 10 towers have unique numerical value. In order to understand why there are 10 such towers, a new counting strategy has to be developed.

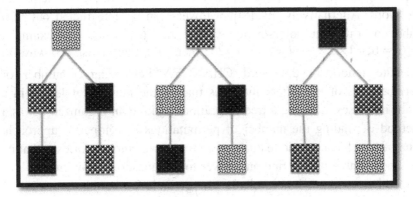

Fig. 2.3. Three-cube towers with non-repeating colors (patterns).

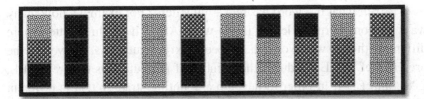

Fig. 2.4. Ten towers out of the total of 27 have unique numerical value.

2.4 Permutation of Letters in a Word

Many real-life situations deal with the concept of permutation – all the possible arrangements of a collection of elements, each of which contains every element once, with two such arrangements differing only in the order of their elements. Commonly, the symbol $P(n)$ denotes the number of permutations of n elements. In this section, the following type of questions will be explored: *How many ways can one permute letters in a word?* To begin, consider possible permutations of letters in the word CAT. This query is similar to the problem of finding the number of different three-cube rods made out of three cubes of different colors/patterns. Indeed, each of the three letters can be associated with one of the three colors/patterns, and thereby, the tree diagram of Fig. 2.3 can be used to represent the permutation of letters in the word CAT.

Alternatively, one can make an organized list showing six permutations of letters in this word: CAT, CTA, ACT, ATC, TAC, TCA.

So, in the case when all letters in an n-letter word are different, according to the rule of product, there exist $n!$ ways to arrange letters in the word. Alternatively, $n!$ distinct towers can be constructed out of n cubes of different colors/patterns, that is, $P(n) = n!$. For example, $P(6) = 6! = 1 \times 2 \times 3 \times 4 \times 5 \times 6 = 720$; therefore, there exist 720 ways to permute letters in the word CHANGE. Such a large number of permutations of six letters indicates that as the number of letters in a word increases, drawing a tree diagram or making an organized list as a method of finding the number of permutations is a hopeless approach. One of the characteristic features of enumerative combinatorics is that its methods enable one to find an answer to the question "How many?" (of something) without listing all the things included in this 'something'. That is why combinatorics is often referred to as the art of counting without counting.

But what if a word has repeating letters? For example, how many ways can one permute letters in the word ALL? If all three letters were different, the answer would have been six permutations. However, the presence of two L's reduces the number of permutations to three: ALL, LAL, and LLA. In order to develop a counting strategy that is different from making an organized list, one can first draw two tree diagrams – iconic (Fig. 2.5) and symbolic (Fig. 2.6) – for the problem of permuting letters in the word ALL.

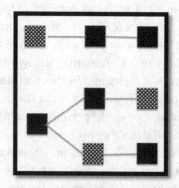

Fig. 2.5. Permutation of letters in the word ALL through tiles.

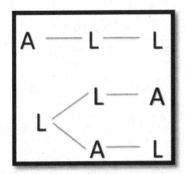

Fig. 2.6. Permutation of letters in the word ALL through letters.

Fig. 2.6 shows that in the word ALL, there are 3 permutations of the letters; yet, one can use Fig. 2.3 to conclude that in the word OLD there are 6 permutations of the letters. Comparing iconic tree diagrams of Fig. 2.3 and Fig. 2.5, one can see that when the stripe pattern is replaced by the mix pattern, each path on the tree is repeated twice because the two mix-patterned cubes are indistinguishable. In that way, in order to count the number of permutations of letters in the word ALL, one has to divide 2! into 3!, as, due to the rule of product, the symbol 2! represents the number of permutations of two objects. In doing so, one gets

$$\frac{3!}{2!} = \frac{1 \times 2 \times 3}{1 \times 2} = 3.$$

This way of counting permutations in a word with repeating letters can also be explained as follows. Let the number of permutations of letters in the word ALL be equal to P. For each such permutation, two L's can be permuted in 2! ways. By the rule of product, the number of permutations of letters in a 3-letter word is equal to $P \times 2!$. On the other hand, the number of permutations in a 3-letter word is equal to 3!. This yields the equation $P \times 2! = 3!$ whence $P = \dfrac{3!}{2!} = 3$.

The following words can be used to illustrate the strategy of counting permutations of letters in words with repeating letters. For example, in the word DOLL the letters can be arranged in

$$\frac{\overset{\text{letters total}}{\overbrace{4!}}}{\underset{\text{two } L\text{'s}}{\underbrace{2!}}} = \frac{1 \times 2 \times 3 \times 4}{1 \times 2} = 12$$ ways; in the word GAGA the letters can be

arranged in $\dfrac{\overset{letters\ total}{\overbrace{4!}}}{\underbrace{(2!)}_{two\ G's} \times \underbrace{(2!)}_{two\ A's}} = \dfrac{1 \times 2 \times 3 \times 4}{(1 \times 2) \times (1 \times 2)} = 6$ ways; in the word

SETTER   the   letters   can   be   arranged   in

$\dfrac{\overset{letters\ total}{\overbrace{6!}}}{\underbrace{(2!)}_{two\ E's} \times \underbrace{(2!)}_{two\ T's}} = \dfrac{1 \times 2 \times 3 \times 4 \times 5 \times 6}{(1 \times 2) \times (1 \times 2)} = 180$; in the word MISSISSIPPI the

letters can be arranged in $\dfrac{\overset{letters\ total}{\overbrace{11!}}}{\underbrace{(4!)}_{four\ I's} \times \underbrace{(4!)}_{four\ S's} \times \underbrace{(2!)}_{two\ P's}} = \dfrac{1 \times 2 \times 3 ... \times 11}{24 \times 24 \times 2} = 34,650$

ways.

As it is always the case in the elementary mathematics, a transition from a small number of objects to a large number of objects to deal with motivates the need for theory that in the case of permutations offers formulas involving factorials of numbers describing those objects. Yet conceptual understanding of the formulas develops in the context of a small number of objects. One way of connecting procedural and conceptual knowledge is through posing a problem [Abramovich, 2015]. In the case of counting permutations one can start using short words and find the number of permutation of their letters using different means – tree diagrams, organized lists, and formulas. Then one can use lengthy words in which case only formulas work.

2.5 Combinations without Repetitions

Modeling of many real-life situations deals with the concept of combination – the selection of a certain number of objects from a given set without regard to their order. Whereas the concept of permutation focuses on the ordering of a certain number of objects (e.g., sorting books *on* a bookshelf), the concept of combination focuses on the selection of a subset of a certain number of objects (e.g., selecting books *from* a bookshelf). Consider the following problem which illustrates the concept of combination.

A Library Problem. *How many ways can one check out four books out of six different library books?*

The chart of Fig. 2.7 shows that if the letters Y and N mean, respectively, "yes" (checking out a book) and "nay" (not checking out a book), then the number of ways four books can be checked out from a six-book collection on display is equal to the number of permutations of letters in the word YYYYNN (only five such permutations are shown). This is a six-letter word with four Y's and two N's. Therefore, the number of permutations sought can be calculated as follows:

$$\frac{\overbrace{6!}^{\text{letters total}}}{\underbrace{(4!)}_{\text{four Y's}} \times \underbrace{(2!)}_{\text{two N's}}} = \frac{1 \times 2 \times 3 \times 4 \times 5 \times 6}{(1 \times 2 \times 3 \times 4) \times (1 \times 2)} = 15.$$

Note that a six-letter word YYNNNN can be used to model the situation of checking out two books out of six books by different authors. Indeed,

$$\frac{\overbrace{6!}^{\text{letters total}}}{\underbrace{(4!)}_{\text{four N's}} \times \underbrace{(2!)}_{\text{two Y's}}} = 15 \quad \text{for the word YYNNNN. Put another way,}$$

$C_6^4 = \dfrac{6!}{4!(5-4)!} = \dfrac{6!}{2!(6-2)!} = C_6^2 = 15$. In general, denoting C_n^m the number of ways to select m objects out of n objects, $n \geq m$, one can write

$$C_n^m = C_n^{n-m} = \frac{n!}{m!(n-m)!}. \tag{2.1}$$

Book 1	Book 2	Book 3	Book 4	Book 5	Book 6
Y	Y	Y	Y	N	N
Y	N	Y	N	Y	Y
N	Y	N	Y	Y	Y
N	Y	Y	Y	Y	N

Fig. 2.7. Library problem as permutation of letters in a word.

Note that as the complexity of mathematical content increases, the diversity of teaching methods associated with the content naturally unfolds. With this in mind, the rule of sum can also be used to find the number of ways to select two books out of six books. To this end, one can first select Book 1 and pair it with each of the remaining five books (Fig. 2.8), something that can be done in $C_5^1 = 5$ ways. In that way, all five pairs that include Book 1 have been formed. In order to form pairs that do not include Book 1, one can select two books out of five books (Fig. 2.9) and this is equal to the number of permutations of letters in the word YYNNN; that is, $C_5^2 = 10$ ways. By the rule of sum we have $C_6^2 = C_5^1 + C_5^2$. In general,

$$C_n^m = C_{n-1}^{m-1} + C_{n-1}^m. \tag{2.2}$$

As an exercise, one can use formula (2.1) to verify the correctness of formula (2.2).

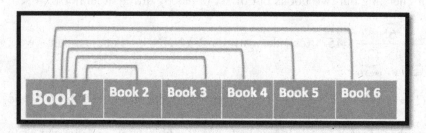

Fig. 2.8. Paring Book 1 with other five books.

Book 2	Book 3	Book 4	Book 5	Book 6
Y	Y	N	N	N
N	Y	N	N	Y
Y	N	Y	Y	N
N	N	Y	Y	N

Fig. 2.9. Selecting two books out of five books.

However, not all methods associated with a given content are readily applicable to each part of the content. That is, the diversity of methods has to be critically evaluated in order to avoid a negative affordance of a generally positive didactic outgrowth. For example, quite frequently, teacher candidates suggest solving the library problem by using a tree diagram. The shortcoming of such an approach is two-fold. First, the increase of the number of books by just one book would significantly increase the size of a tree diagram. Second, a tree diagram would display the arrangements of books in different orders rather than their combinations (when the order of books in a combination is immaterial). In other words, combinations don't follow the rule of product.

2.6 Combinations with Repetitions

The best way to introduce a new concept is to begin with a situation that can be resolved by either using a pre-requisite knowledge or constructing a physical model that represents the situation. Then, drawing on the limitations of the physical model (experiment), one can motivate the concept, which, once developed theoretically, can be verified experimentally. To this end, consider

A Dunkin' Donuts Problem. *How many ways can one buy two donuts out of three types: Chocolate, Glazed and Jelly?*

This new situation is close to combinations since the order in which the donuts are selected is irrelevant. However, the difference between checking out library books and purchasing donuts is that whereas one can buy two donuts of the same type (assuming their availability), two copies of the same book (even if they exist) may not be checked out. Let us investigate how the change of context affects the model used to describe this context in mathematical terms. To begin, we create an experiential model of the situation as shown in Fig. 2.10. The value of an experiential model using a small number of objects is that the model can be used later as a verification of a theoretical tool in that particular case. This is consistent with the point of view expressed by Gestalt psychologists Luchins and Luchins [1970a] arguing for the merit of intuition in developing mathematical understanding and fostering deductive reasoning: "it seems that when the intuitive approach is used in conjunction (but not as a substitute for) the deductive proofs, it results in clearer understanding and better retention of the proofs as well as wider application of the formulas" (p. 268). Gestalt psychology [Ellis, 1938], as will be shown in Chapter 7, is a highly mathematically oriented theory.

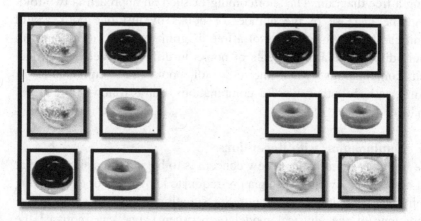

Fig. 2.10. Selecting two donuts out of three types.

Fig. 2.10 shows that two donuts out of three types can be selected in six ways. Now, we have to develop a model and offer a theory-oriented counting tool confirming the experimental finding displayed in Fig. 2.10. Unfortunately, what can also be seen from the chart of Fig. 2.11 is that using the "Y/N" model results in the variation of the number of letters in a word as one moves from one selection of two donuts to another. For example, whereas the selection of any two different donuts is represented by a three-letter word YYN, the selection of, say, two Chocolate donuts, is represented by a four-letter word YYNN. This suggests that the old model used to describe combinations without repetitions in the context of checking out library books cannot be used for the description of combinations with repetitions in the context of buying donuts. So, the change of real-life context requires the change of theoretical model.

The deficiency of the "Y/N" model that was described above motivates the development of a new model for the Dunkin' Donuts problem. Yet, this new model should be capable of preserving the main idea of using the technique of permuting letters in a word. A possible solution, which does require insight [Dunker, 1945] alternatively referred to as productive thinking [Wertheimer, 1959] or displacement of concepts [Schon, 1963], is not to use the letter N (replacing a variable characteristic by an invariant one) in a word because the number of N's, as we have observed, depends on a selection. A characteristic that is invariant across all selections of two donuts is the number of donut types. With this in mind, the letter B can be introduced to serve as a *boundary* between any two types of donuts. Such a modified chart is pictured in Fig. 2.12 where, for instance, the word YBYB describes the selection of Chocolate and Glazed donuts. In that way, a four-letter word with two Y's and two B's describes a possible selection of two donuts out of three types. The total number of such selections (often referred to as combinations with repetitions) is equal to $\dfrac{4!}{(2!) \times (2!)} = 6$ – the number of permutations of the letters in the word YYBB. This answer, obtained through the "Y/B" model, coincides with the experimental result shown in Fig. 2.10.

Fig. 2.11. The deficiency of the "Y/N" model.

Note that whereas selecting two donuts out of three types can be done through making an organized list, only the "Y/B" model can answer the question about selecting a larger number of donuts, say, ten, from three types. The corresponding "Y/B" model is represented by a word with ten Y's and two B's; the number of permutations of the letters in such a word is equal to $\dfrac{12!}{10! \times 2!} = 66$. In general, denoting $\overline{C}_n^m$ the number of n-combination with repetitions out of m types, we have

$$\overline{C}_n^m = \frac{[n+(m-1)]!}{n!(m-1)!}. \tag{2.3}$$

⬤		🍩		🍩
Y	B	Y	B	
	B	Y	B	Y
Y	B		B	Y
YY	B		B	
	B	YY	B	
	B		B	YY

Fig. 2.12. The "Y/B" model.

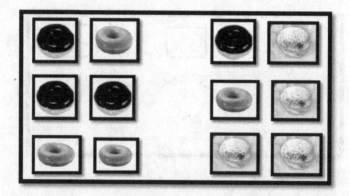

Fig. 2.13. Selecting two donuts out of three types through the rule of sum.

The problem of selecting two donuts out of three types can be also solved by using the rule of sum. Once again, the complexity of mathematical content calls for the diversity of methods of teaching the content. To this end, first, one make all selections of two donuts out of two types, say, from Chocolate and Glazed (the left part of Fig. 2.13) –

this can be done in $\overline{C}_2^2 = \dfrac{[2+(2-1)]!}{2! \times (2-1)!} = 3$ ways. Alternatively, the number of such selections is equal to the number of ways that the letters in the word YYB can be permuted, that is, in $3!/2! = 3$ ways. Then, one can select one donut out of the three types, that is, Chocolate, Glazed, and Jelly and add as the second donut a Jelly one not used in the first selection (the right part of Fig. 2.13) – this can be done in $\overline{C}_1^3 = \dfrac{[1+(3-1)]!}{1! \times (3-1)!} = 3$ ways. Alternatively, the number of such selections is equal to the number of ways that the letters in the word YBB can be permuted, that is, in $3!/2! = 3$ ways. Thus in the case $n = 2$ and $m = 3$ the rule of sum yields $\overline{C}_2^3 = \overline{C}_2^2 + \overline{C}_1^3$. In general,

$$\overline{C}_n^m = \overline{C}_n^{m-1} + \overline{C}_{n-1}^m. \qquad (2.4)$$

As an exercise, one can use formula (2.3) to prove the correctness of formula (2.4).

Finally, the "Y/B" model can be applied to the problem of finding all possible numerical values represented by three-cube towers constructed out of three colors each of which is assigned one of the numbers – 1, 5, and 10. The number of such values is equal to the number of permutations of letters in the word YYYBB (three cubes and two boundaries for three colors), that is, $\dfrac{5!}{3! \times 2!} = \dfrac{3! \cdot 4 \cdot 5}{3! \times 2} = 10$.

To conclude, note that counting techniques of this chapter were based on formulas involving the products of consecutive natural numbers starting from one. A practice of using these formulas requires conceptual understanding that not all multiplications have to be (or even may be) carried out in a straightforward way because of the fastest growth of the factorial function. Indeed, already 13! is a ten-digit number. The fractional character of the formulas makes it possible to reduce the magnitude of the resulting products because the number of permutations of objects that they describe decreases as the number of repetitions of objects increases. Such conceptualization is reflected in a possibility to cancel common factors (products) in the specific formulas. It is important that teacher candidates appreciate this possibility and can explain it conceptually using the structure of models that are used as means of resolving counting situations.

Chapter 3

Counting and Reasoning with Manipulative Materials

3.1 Introduction

This chapter demonstrates various uses of manipulative materials in the context of teaching elementary mathematics to prospective schoolteachers. It addresses a number of notable recommendations for mathematics teacher preparation as well as several standards for teaching primary school mathematics where the integration of tactility and abstraction is the main methodological focus of instruction. The chapter illuminates a didactic approach to teaching mathematics through modeling with manipulative materials by creating isomorphic relationships between the first and the second-order symbols as defined, in the context of teaching language, by Lev Vygotsky – a Russian educational psychologist of the 20th century – whose work greatly influenced the development of psychology and views on education in the United States. Using the words by Vygotsky [1978], one can say that mathematical knowledge had been developed through the transition from dealing with the "first-order symbols ... directly denoting objects or actions ...[to] the second-order symbolism, which involves the creation of written signs for the spoken symbols of words" (p. 115). Put another way, the approach to teaching mathematics through modeling with manipulative materials is based on the de-contextualization of hands-on activities requiring just intuition and common sense. As noted in the Common Core State Standards [2010], "Mathematically proficient students ... bring two complementary abilities to bear on problems involving quantitative relationships: the ability to *decontextualize* – to abstract a given situation and represent it symbolically ... and the ability to *contextualize* ... [that is] to probe into the referents for the symbols involved" (p. 6, italics in the original). The approach is grounded into a pedagogical tradition that views children building formal knowledge on the foundations of their abilities that they use in an informal, intuitive manner. In terms of language acquisition, it was suggested that the role of formal schooling is to make a student learn "conscious awareness of what he does … [that is, the student] learns to operate on the foundation of his capacities in a volitional manner. His capacity moves from an unconscious, automatic plane to a voluntary, intentional, and conscious plane. Instruction in written speech and grammar play a fundamental role in this process" [Vygotsky, 1987, p. 206].

Indeed, a 3-year old child is capable of saying "I saw a cat in the backyard" but would be puzzled when asked using the verb "see" in the past tense. At the same time, an adult learner of English is not capable of saying the same phrase without learning English grammar first. Similarly, the teaching of formal mathematics should capitalize on learners' intuitive ownership of mathematical ideas and techniques that in many cases stem from common sense. For example, when asked who is taller, Jack (a first grader) or John (a six grader), a child would immediately say "John". How does the child know that? Most likely by seeing that a part of John's body is above Jack's body; in other words, the child intuitively uses the concept of difference as a tool in comparing the boys' heights. However, when given these heights in inches or centimeters, the child, without guessing, would (most likely) not be able to give a correct response.

A psychological explanation of this comparison can be found in the pioneering research concerning the relationship between visual imagery and deductive reasoning [DeSoto *et al.*, 1965]. These authors argued that because the relations of *greater* and *smaller* have a fixed link to the vertical axis in human cognitive space, a subject who is told that A is *greater* (*smaller*) than B forms an image of A *above* (*below*) B. Likewise, seeing John's head above Jack's head makes it possible to decide who is taller. This, along with more recent studies on the relationship between visual images and spatial representations [Knauff and May, 2006], imply that learners are good at thinking of objects as spatially ordered. That is why, the current emphasis of the Common Core State Standards [2010] on the use of pictures and diagrams in problem solving helps create the foundation for using intuitive understanding prior to learning formal mathematics: "Mathematically proficient students … can construct arguments using concrete referents … even though they [arguments] are not generalized or made formal until later grades … [and to] identify important quantities in a practical situation and map their relationships using such tools as diagrams, two-way tables, graphs, flowcharts and formulas" (p. 7). Similarly, Ministry of Education Singapore [2012], referring to the skills needed for problem solving, suggests, "to make sense of various mathematical ideas … students should be exposed to … hands-on activities … to help them relate abstract mathematical concepts with concrete experiences" (p. 15). In much the same spirit, the Conference Board of the Mathematical Sciences [2012] recommends, "In order to … help students use [concrete technological tools] strategically in doing mathematics, teachers need to understand the mathematical aspects of these tools and their uses" (p. 2) and, in addition, the teachers need "to develop the ability to critically evaluate the affordances and limitations of a given tool, both for their own learning and to support the learning of their students" (p. 34). That is, the diversity of teaching methods provided by the diversity of available teaching tools should not be taken to mean that the tactility and visual appeal of the first-order symbols enable their accurate de-contextualization at the level of the second-order symbolism regardless of the mathematical content involved.

3.2 Constructing a Triangle out of Straws

One of the activities found in a curriculum guide of New York State Education Department [1998] recommends that students in grades 1 and 2 use straws to build geometric figures. As presented, this hands-on activity is in support of children's mathematical thinking and reasoning skills. Typically, straws available for this exploration are all the same size and any plane figure – a triangle, a quadrilateral, a pentagon, and so on, except, perhaps, circle, can be constructed out of straws. When this activity is discussed within a mathematics teacher education course, mathematically unsophisticated teacher candidates could perceive such use of straws as a "really funny" way of demonstrating how a particular figure looks like and, thereby, their joy about mathematics might end after a figure is constructed. On the contrary, the task with straws has the potential to extend this amateurish sense of a learning experience to enable, in the spirit of the experimental mathematics approach [Arnold, 2015; Abramovich, 2014; Boas, 1971; Borwein and Bailey, 2004;

Sutherland, 1994], the discovery of knowledge, which is unlikely to be found in any part of school mathematics curriculum.

What teacher candidates are supposed to learn in the modern university classroom is how to go beyond the seemingly mundane character of this (or similar) task they naively consider a 'worthwhile exploration' simply due to its trendy hands-on setting and the use of everyday objects as a mathematical manipulative. Nonetheless, such an activity might end immediately after a figure is constructed unless a teacher knows how to extend it towards the development of students' mathematical reasoning skills. At the very least, as a reflection on the activity, some simple questions can be asked: How many straws does one need to construct a triangle, a square, a pentagon? Assuming that one typically uses three straws for triangle, four straws for square, five straws for pentagon, such a query follows the framework "single question – single answer", not allowing for the growth in mathematical thinking. In order to overcome the limitations of this framework, the students should be encouraged to demonstrate multiple ways of constructing a particular geometric figure. Such kind of encouragement to think deeper about mathematics follows the idea of "recognizing in a result something that can be turned into a question" [Mason, 2000, p. 98]. For example, after triangle and square have been constructed out of, respectively, three and four straws, this modest result may be turned into a question "posed in a routine way so as to obscure mathematical thinking" [*ibid*, p. 103]. For example, a teacher may ask: How can one use four, five, six, and so on, (same size) straws to construct a triangle, a square (or any quadrilateral), and a pentagon?

As one can see, a teaching method used here is asking a question that does not presuppose a single answer. One can also see here a limitation of the same size straws in making certain constructions. After some explorations with the straws, students can see that one cannot construct a triangle out of four such straws but can do that with five and six straws. A mathematical concept underlying these activities with straws is the so-called triangle inequality – the sum of any two sides of a triangle is greater than the third side – a profound concept that everybody intuitively possesses without even realizing it.

To explain the last statement, consider the case when a student is asked to cut a straw into three pieces out of which a triangle has to be constructed. Without knowing the triangle inequality, the student (most likely) would have the longest piece shorter than the combined length of the other two pieces. Likewise, children (and adults alike) often cut across the grass rather than walk along the pavement to get to the building faster. Nonetheless, without appropriate intervention, a student could believe that one can always construct a triangle out of any three straws, equal or unequal in size. Such intervention is very important for one's conceptual development and growth in mathematical thinking. So, unless a counter-example is provided, the triangle inequality is unlikely to be discovered; that is, a switch from intuitive knowledge to conceptual knowledge is not automatic.

Fig. 3.1. A counterintuitive (unlikely) cut of a straw.

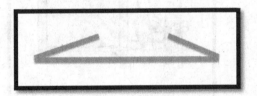

Fig. 3.2. Attempting conceptualization of the unlikely cut,

To this end, students can first be asked to construct a triangle out of four equal straws. Alternatively, they can be presented with three pieces like those shown in Fig. 3.1 and asked to construct a triangle using entire pieces as the side lengths of a triangle. In both cases, triangle construction turns out being impossible. Why is it so? This question seeks an explanation of the observed phenomenon provided by a counter-example. When trying to make the endpoints of the two shorter straws meet by moving them closer and closer to the longest straw (Fig. 3.2), the relationship among the lengths of the straws may become detectable: the combined lengths of the first two, by being smaller than that of the longest straw is what prevents us from constructing a triangle. Therefore, the largest side length must be smaller than the sum of the other two side lengths. This is the core of the triangle inequality.

A similar manifestation of intuitive mathematical knowledge can transpire during a lesson on measurement in grade two. As observed by the author, when measuring distance from a mark on the floor to the wall (Fig. 3.3), young children were placing a measuring tape perpendicular to the wall without any formal knowledge of mathematics behind this measuring activity. Nonetheless, unlike the case of the triangle inequality, it is only through measuring different distances from the point to the line that this fact can be confirmed without using mathematical machinery way beyond the elementary level.

Fig. 3.3. The shortest distance as an intuitive concept.

3.2.1 *Reflection on the activity with straws*

The triangle inequality can be formulated in at least two different ways. The first formulation is that the sum of *any* two side lengths should be greater than the third side length. For example, for the triple of integers 3, 5, and 10 we have the inequality $3 + 10 > 5$, yet triangle cannot be constructed out of the straws measured 3, 5, and 10 linear units. Why is it so? It is because of the presence of the word *any* in the above formulation. We see that $3 + 10 > 5$ and $5 + 10 > 3$ yet $3 + 5 < 10$ and therefore, not *any* two side lengths are greater than the third side. This discussion is helpful for understanding the meaning of a mathematical definition. Put another way, the inequality $3 + 5 < 10$ serves as a counter-example to the statement of the triangle inequality. Another formulation of the triangle inequality – the longest side length must be smaller than the sum of other two sides – directly points to the only one inequality to verify which one has to select the longest side and compare it to the sum of the other two sides.

Let us compare activities that are expected in order to check out the formulations (definitions). In the first case, given the triple (a, b, c), one has to construct all possible two-element sums: $a + b, a + c, b + c$, and then compare them to, respectively, c, b, a. That is, three sums have to be compared to three numbers. In all, one has to carry out three additions and three comparisons. In the second case, the largest element has to be selected from the triple (a, b, c) and then compared to the sum of the remaining two elements. Here, one has to carry out three comparisons and one addition. One may ask: Why do we need the first definition when its verification requires two extra additions? In the first case, we try all three cases to check whether a counter-example could be created. Alternatively, one can arrange the three elements from the least to the greatest (this requires two comparisons), add the first two, compare their sum to the third element to see whether the third comparison offers a counter-example. This alternative is equivalent to the verification of the second case, when one selects the only possible case and sees whether it offers a counter-example.

Another discussion may deal with a qualitative explanation of intuitive actions, that is, cutting a straw, in the formal language of mathematics. Suppose the elements of the triple (a, b, c) are arranged from the greatest to the least, that is, $a > b$ and $b > c$. Now, compare a to $b + c$. The result depends on the relationship between $a - b$ and c. If $a - b > c$ then $a > b + c$; if $a - b < c$ then $a < b + c$. In other words, in order for the triangle inequality to be satisfied, both b and c should not be too small or b should be rather large in order to make the difference $a - b$ smaller than c. All these considerations find their intuitive recognition when one cuts a straw into three pieces out of which a triangle can be constructed or one cuts across the grass instead of walking along the pavement to get to the building faster.

3.2.2 *A real-life application of the triangle inequality*

A real-life application of the triangle inequality and teaching methodology stemming from the application can be demonstrated through the following situation. Suppose you have to buy at the lumberyard three wooden sticks to make a garden bed shaped in the form of triangle. A sales associate offers you the following three pieces: a 10, 5, and 3-foot long sticks, priced by the number of feet. If you like the sticks but do not know the triangle inequality, you would take the offer only to discover later, at home, that one of the sticks is too long to be used as a side of a triangle; that is, you have paid extra money. If you know the triangle inequality, then you would have to do the following (basic) mathematical actions with the three integers: comparison and addition. That is, among the integers 10, 5 and 3 you have to select the largest one, 10; then add the remaining two integers, 5 and 3, to get 8; then compare 8 to 10, and conclude that because 10 is greater than 8, a triangle cannot be constructed. (At the very basic level, the comparison of the number 10 to the sum 5 + 3 resulting in the inequality 10 > 8 can be made using the first-order symbols as shown in Fig. 3.4 or using the second-order symbols by finding that the difference 10 − 8 is a positive number; that is, 10 is, indeed, greater than 8). After you reported your mathematical findings to the sales associate, the latter said it is possible to make the longest stick shorter by cutting off pieces measured by the whole number of feet. How can this be done?

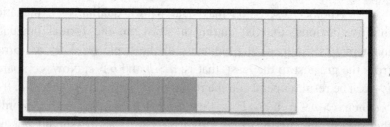

Fig. 3.4. Comparing 10 to 5 + 3 through the first-order symbols.

Here we have another opportunity to integrate the framework "single question – multiple answers" as a teaching method. This situation includes an application of cognitive skills requiring one to subtract, select, add, compare, and conclude (S^2AC^2). Let us make the stick one foot shorter and see what would happen. We begin with subtraction: $10 - 1 = 9$. Then, select from the triple (9, 5, 3) the largest number, 9; add $5 + 3 = 8$, compare 9 and 8, and conclude that because $9 > 8$ a triangle cannot be constructed. Therefore, one has to decrease 10 by another foot and apply the S^2AC^2 algorithm. Now we subtract $10 - 2 = 8$, select from the triple (8, 5, 3) the largest, 8; add $5 + 3 = 8$; compare 8 to 8, and conclude that because $8 = 8$ a triangle cannot be constructed. Therefore, one has to decrease 10 by yet another foot and apply the S^2AC^2 algorithm. Now we subtract $10 - 3 = 7$, select from the triple (7, 5, 3) the largest number, 7, add $5 + 3 = 8$, compare 8 to 7 and conclude that because $7 < 8$, the triangle inequality $7 < 5 + 3$ is satisfied and a triangle with the side lengths 7, 5, and 3 can be constructed. However, another triangle is possible, if we subtract $10 - 4 = 6$, select 6 from the triple (6, 5, 3), add $5 + 3 = 8$, compare 8 to 7 and conclude that because $6 < 5 + 3$, a triangle with the side lengths 6, 5, and 3 may be constructed. Apparently, this is not the last triangle that can be constructed by decreasing 10 by a whole number of feet. But the next subtraction yields a triple which makes it problematic to apply the second step of the S^2AC^2 algorithm.

3.2.3 *Modifying the S^2AC^2 algorithm to enable linguistic coherency*
Indeed, a slightly different situation can be observed when 10 is decreased by 5 to become 5. Now, one has to select the largest number from the triple (5, 5, 3). Two equal numbers are the elements of the triple and either one may be selected as the largest. A pupil may ask: How come? There is no easy answer to this question. In fact, the largest number does not exist, just as there is no single winner of a tight game. A possible way to avoid answering this question is to go back and revise the S^2AC^2 algorithm: arrange three integers from the greatest to the least allowing for either two or all three of them to be equal and then compare the first number to the sum of the other two numbers. That is, in the current case, we have the arrangement 5, 5, 3, and then compare the first (far left) number to the sum of other two numbers, $5 < 5 + 3$, to conclude that the triple (5, 5, 3) may serve as the side lengths of a triangle. It may be up to a teacher to decide whether to modify the S^2AC^2 algorithm to become A^2C^2 (**arrange-add-compare-conclude**) once a triple with no

largest element comes to existence or to suggest the latter algorithm at the beginning of the activities. In fact, difficult questions may be asked in either case and until a special case emerges, there might be no need to talk about two or three equal side lengths.

With the A^2C^2 algorithm, one only lines up three numbers not allowing any number to be smaller than the one immediately to its right. As an aside, note that such an algorithm may be helpful for young children's understanding of how to order numbers when there are equal numbers. In fact, the A^2C^2 algorithm works perfectly well for the triples obtained through the reduction of 10 by 3 and 4 (section 3.2.2). Note that the described transition from one algorithm to another is an example of how a slight change of content calls for a new teaching method that is invariant to this change.

In this regard, the following field observation is worth noting. A pre-student teacher during her internship in a kindergarten classroom gave each pupil a handful of colored square tiles, grid paper, and crayons asking them to arrange tiles by colors in the form of a bar graph, draw its replica on the paper, and report the color they have the most and the color they have the least. Several pupils soon raised hands and looked confused. It turned out that they had the repeating number of colors in either the most or the least categories (or both like in Fig. 3.5). The intern had difficulty revising the task on the fly and became upset. The author discussed the situation with her and suggested that the pupils could have been asked to describe their drawing using just one more term: *the same*. Another possibility to avoid confusion could be to ask the pupils to arrange bars in such a way that no bar may be smaller than the one immediately to its right and then describe their graph using the words *most*, *least*, and *the same*.

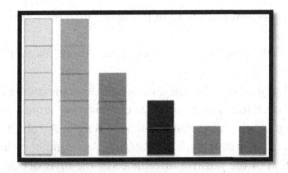

Fig. 3.5. A confusing question: Which color do you have the most/the least?

3.2.4 *How many triangles can be constructed?*

The A^2C^2 algorithm can be applied to the case when 10 is reduced by 6 to yield the triple (5, 4, 3) for which the triangle inequality is satisfied as $5 < 4 + 3$, thus providing an option for another triangular shaped garden bed. Reducing 10 by 7 yields (5, 3, 3) with $5 < 3 + 3$. Further reduction of 10 yields (5, 3, 2) and the equality $5 = 3 + 2$, once again, defies the triangle inequality. Therefore, a single question about how a 10-foot stick can be modified to match the construction of a triangle yields five answers, namely, (7, 5, 3), (6, 5, 3), (5, 5, 3), (5, 4, 3), and (5, 3, 3). Finally, through this discussion one can ask: Is any such triangle unique? For example, can one construct two different triangles with the side lengths, given by the triple (5, 4, 3)? Why or why not? How can this situation be explored? Encouraging learners of mathematics to ask questions as a teaching method opens a window to a new mathematical content. In particular, through exploring the triangle inequality in an open way, one can learn about the side-side-side postulate of congruency of triangles through a hands-on activity.

Another worthwhile extension of the construction of triangles using straws is to find the number of integer-sided triangles the side lengths of which do not exceed a given number. This task, requiring an ability to de-contextualize or change a context for the sake of hands-on convenience, can be formulated in terms of towers built out of square tiles as follows.

Construct all possible tri-towers satisfying the following two conditions:

(i) no tower to the right is smaller than the tower immediately to the left, and

(ii) the first two towers when placed one above the other make a tower taller than the third one.

Put another way, the essence of de-contextualization is that the height of each tower in a tri-tower combination represents a side length of a triangle with a tile being the unit of length. An ability to move from one type of manipulative materials to another indicates one's appreciation of the positive affordance of the diversity of methods made possible by the diversity of tools. Fig. 3.6 shows that when the third tower is four-tile tall, there exist six such towers; in other words, there exist six integer-sided triangles with the side lengths

$(1, 4, 4), (2, 4, 4), (3, 4, 4), (2, 3, 4), (3, 3, 4),$ and $(4, 4, 4)$.

Furthermore, one can explore a change in the number of triangles when one makes a transition from four to five. In this case, the third element in the above six triples will become five yielding the new set of triples

$(1, 4, 5), (2, 4, 5), (3, 4, 5), (2, 3, 5), (3, 3, 5),$ and $(4, 4, 5)$

out of which two triples do not satisfy condition (ii), namely, $(1, 4, 5)$ and $(2, 3, 5)$. The corresponding pair of tri-towers is shown in the top-left corner of Fig. 3.7. The elements of the disappearing triples satisfy the relations $1 + 4 = 5$ and $2 + 3 = 5$. That is, the number of these triples equals to the number of non-ordered partitions of the largest element in two summands. By the same token, five new tri-towers can be constructed each of which has two towers of the same height, 5. Numerically, they are represented by the triples $(1, 5, 5), (2, 5, 5), (3, 5, 5), (4, 5, 5)$ and $(5, 5, 5)$, the first element of which numerates the new triples. In general, the total augmentation of the number of triangles through the transition from the largest side n to the side $n + 1$ is equal to $n - INT(n/2)$ where the subtrahend, called the greatest integer function and often tagged INT, returning the largest integer smaller than or equal to $n/2$, defines the number of non-ordered partitions of integer n in two summands (Chapter 1, section 1.2). For example, when $n = 5$ we have $INT(5/2) = 2$ and $5 = 1 + 4 = 2 + 3$; when $n = 6$ we have $INT(6/2) = 3$ and $6 = 1 + 5 = 2 + 4 = 3 + 3$.

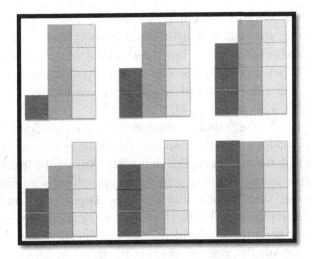

Fig. 3.6. A manipulative task with a hidden meaning.

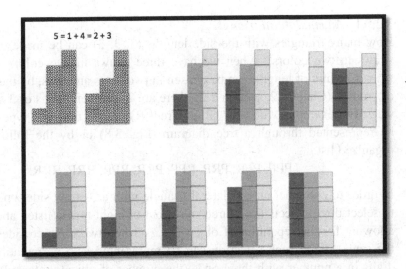

Fig. 3.7. Transition from four to five.

3.2.5 *Using multicolored straws*

What if we have straws in different colors? Such straws may be used in the classroom to make mathematics even more "akin to that of playing games … [seen by educators as] the spontaneous way in which children acquire much of their mastery over the environment" [Gattegno, 1963, pp. 80–81]. With this in mind, consider the case of two colors with a straw being the unit of length. *How many different triangles can be constructed out of straws in two colors if the longest side comprises three straws?* When straws are the same size and color, there exist four such triangles the side lengths of which are given by the triples (3, 3, 3), (3, 3, 2), (3, 3, 1), and (3, 2, 2).

These four triples represent one equilateral triangle, two isosceles triangles with the base being the smaller side, and one isosceles triangle with the base being the larger side. Among the triples, there is no triple representing a scalene triangle because there is no scalene triangle with integer sides among which 3 is the largest. The smallest scalene triangle is represented by the triple (4, 3, 2). Therefore, the following four cases need to be considered: an equilateral triangle, an isosceles triangle the smaller side of which is the base, an isosceles triangle the larger side of which is the base, and a scalene triangle.

3.2.5.1 *An equilateral triangle*

How many triangles with the side lengths (3, 3, 3) can be made of the straws in two colors? When we have three straws in two colors, each straw (the unit of length) can be chosen in two ways and thus, by the rule of product (Chapter 2, section 2.2), there are 2^3 different types of 3-straw sides. For example, with pink (P) and red (R) straws, the eight types can be represented through a tree diagram (Fig. 3.8) or by the following organized list

PPP, PPR, PRP, RPP, PRR, RPR, RRP, RRR.

In order to construct an equilateral triangle with a 3-straw side, one has to select three objects (i.e., three sides) out of eight types (listed above), allowing for the repetition of objects (e.g., when two or three sides are the same). Each such selection can be represented as a permutation of digits in a number with three ones (the number of objects selected) and seven zeroes (serving as boundaries among the eight types – in the above organized list, commas are used as boundaries); the number of permutations of the digits in the number 1110000000 is equal to $\dfrac{(3+7)!}{3!\,7!} = 120$ (Chapter 2, section 2.6, formula (2.3) – this formula will be used throughout section 3.2.5). For example, among the 120 permutations, the number 1110000000 represents a triangle all sides of which are of the type PPP; the number 1000000101 (another permutation of the ten digits) represents a triangle the sides of which are of the types PPP, RRP, and RRR. An interesting exercise is to associate different permutations of digits in the (ten-digit) number 1110000000 with the types of sides of a triangle.

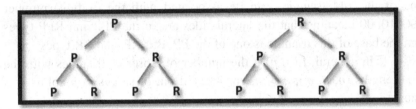

Fig. 3.8. Counting equilateral triangles using a tree diagram.

In general, $E(n, p)$ – the number of equilateral triangles with the side length n constructed out of straws in p colors – is equal to

$$\frac{(3+p^n-1)!}{3!(p^n-1)!} = \frac{(p^n+2)!}{6(p^n-1)!} = \frac{p^n(p^n+1)(p^n+2)}{6}$$

as the number of permutations of digits in the number 11100...0 with three ones (the number of sides in a triangle) and p^n-1 zeroes (serving as boundaries among p^n types of multicolored sides constructed out of n straws in p colors). That is,

$$E(n, p) = \frac{p^n(p^n+1)(p^n+2)}{6}.$$

When $p = 2$ and $n = 3$ we have $\dfrac{2^3(2^3+1)(2^3+2)}{6} = \dfrac{8 \cdot 9 \cdot 10}{6} = 120$

triangles.

3.2.5.2 *An isosceles triangle with the base being the smaller side*

How many triangles with the side lengths (3, 3, 2) can be made of the straws in two colors? Firstly, there are 2^2 different types of 2-straw sides (see the top and the middle levels of the tree diagram of Fig. 3.8). Secondly, for equal sides, we have to select two objects out of 2^3 types (shown in the tree diagram of Fig. 3.8), something that can be done in

$\dfrac{(2+7)!}{2!7!} = 36$ ways – the number of permutations in the (9-digit) number

110000000 with two ones and seven zeroes. For each such selection of a pair of 3-straw sides there are 2^2 selections of a 2-straw side. By the rule of product, there exist $36 \cdot 4 = 144$ isosceles triangles with the side lengths (3, 3, 2) constructed out of straws in two colors. For example, one of the 144 triangles can be associated with the (9-digit) number 100010000 meaning that the lateral sides are of the PPP and RPP types and the base of this triangle is one of the PP, PR, RP, and RR types.

In general, $I_1(n, p)$ – the number of isosceles triangles with the side lengths (n, n, m), $n > m$, constructed out of straws in p colors – is equal to

$$\frac{(2+p^n-1)!}{2!\left(p^n-1\right)!} \cdot p^m = \frac{(p^n+1)!}{2\left(p^n-1\right)!}p^m = \frac{p^{n+m}(p^n+1)}{2}.$$

That is,

$$I_1(n, p) = \frac{p^{n+m}(p^n+1)}{2}.$$

When $p = 2$, $n = 3$, and $m = 2$ we have $\dfrac{2^5(2^3+1)}{2} = \dfrac{32 \cdot 9}{2} = 144$ triangles.

Likewise, when $p = 2$, $n = 3$, and $m = 1$ we have $\dfrac{2^4(2^3+1)}{2} = \dfrac{16 \cdot 9}{2} = 72$

triangles with the side lengths (3, 3, 1) constructed out of straws in two colors.

3.2.5.3 *An isosceles triangle with the base being the larger side*
How many isosceles triangles with the side lengths (3, 2, 2) made of the straws in two colors are there? This time, we have to select two objects

out of 2^2 types, something that can be done in $\dfrac{(2+3)!}{2!3!} = 10$ ways – the

number of permutations of digits in the (5-digit) number 11000. For each such selection, there are 2^3 types of 3-straw sides. Therefore, by the rule of product, there exist 80 ($= 10 \cdot 8$) isosceles triangles with the side lengths (3, 2, 2) constructed out of straws in two colors. For example, one of the 80 triangles can be associated with the number 10001 meaning that the lateral sides are of the PP and RR types and the base of this triangle is one of the eight types listed in section 2.5.1.

In general, $I_2(n, p)$– the number of isosceles triangles with the side lengths (n, m, m), $n > m$, $n < 2m$, constructed out of straws in p colors – is equal to $\dfrac{(2 + p^m - 1)!}{2!(p^m - 1)!} \cdot p^n = \dfrac{(p^m + 1)!}{2(p^m - 1)!} p^n = \dfrac{p^{n+m}(p^m + 1)}{2}$.

That is,

$$I_2(n, p) = \frac{p^{n+m}(p^m + 1)}{2}.$$

When $p = 2$, $n = 3$, and $m = 2$ we have $\dfrac{2^5(2^2 + 1)}{2} = \dfrac{32 \cdot 5}{2} = 80$ triangles with the side lengths $(3, 2, 2)$ constructed out of straws in two colors.

3.2.5.4 *A scalene triangle*

How many scalene triangles with the side lengths $(4, 3, 2)$ can be constructed out of straws in two colors? The largest side can be constructed in $2^4 = 16$ ways, the middle side can be constructed in $2^3 = 8$ ways, and the smallest side can be constructed in $2^2 = 4$ ways. By the rule of product, the number of scalene triangles having the side lengths $(4, 3, 2)$ is equal to $2^4 \cdot 2^3 \cdot 2^2 = 2^{4+3+2} = 2^9 = 512$.

In general, consider the case of constructing triangles with the side lengths (n, m, k), $n > m > k$, $n < m + k$, out of straws in p colors. Each of the sides can be constructed, respectively, in p^n, p^m, and p^k ways. By the rule of product, the number of such triangles is equal to

$$S(n, p) = p^{n+m+k}.$$

When $p = 2$, $n = 4$, $m = 3$, and $k = 2$ we have $S(4, 2) = 2^9 = 512$.

3.3 Two Types of Representation as Means of Transition from Visual to Symbolic

In this section, using Vygotsky's [1978] terminology (already cited in section 3.1), the distinction will be made between the first-order symbols and second-order symbolism, which develops by "shifting from drawing of things to drawing of words" (p. 115). Indeed, three pieces of a straw out of which a triangle can be constructed may be construed as the first-order symbols and the triangle inequality (along with the associated numeric relations) as the second-order symbolism. Likewise, the description of the diagram of Fig. 3.9 made out of two-color counters (typically, having one side red and another side yellow) described at the junior primary level (age 5-6) as an *AB*-pattern (or patterns), is an example of what may be construed as the rudiments of *early algebra*. Indeed, seeing in both lines of the diagram an *AB*-pattern rather than seeing the top one as an *AB*-pattern and the bottom one as a *BA*-pattern (or vice versa) leads to one's appreciation of the concept of variable when the same letter assumes different numerical values. In turn, the language of algebra used in the formal description of manipulative patterns is an example of the second-order symbolism, something that can be taught by revealing hidden meaning of these patterns.

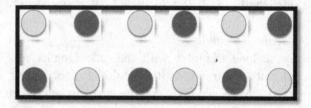

Fig. 3.9. Counters as the first-order symbols.

It is through play that young children can move from creating diagrammatic representations of patterns or more complicated mathematical situations involved to describing them in the language of mathematics. A practical implication of the approach to teaching second-order mathematical symbolism by using manipulative materials as the first-order symbols is twofold: it can be done earlier than usual and also through a more effective pedagogy. Note that Vygotsky [1978] made a similar recommendation emphasizing the role of play in teaching writing and argued, "writing must be "relevant to life"—in the same way that we require "relevant" arithmetic" (p. 118). The construction of triangles out of straws as described above is an example of "relevant mathematics" when the application of arithmetic to geometry renders the integration of two curricula strands. Indeed, deciding if a triple of numbers can represent side lengths of a triangle is an example of relevant (or situated) arithmetic when the S^2AC^2 or A^2C^2 algorithms are performed on integers in the context of geometry. In particular, it shows how arithmetic can be used as a tool in real-life geometric applications.

3.4 Signature Pedagogy of Elementary Mathematics Teacher Education

One can say that teaching mathematics in the elementary school by connecting first-order symbols and second-order symbolism, belongs to a deep structure of teacher education *signature pedagogy* defined as "a set of assumptions about how best to impart a certain body of knowledge and know-how" [Shulman, 2005, p. 55]. A well-known observation that teachers use to teach by replicating ways they have been taught reinforces the significance of such assumptions. Indeed, the way teachers teach shapes their professional behavior and at the time when "a society [is] so dependent on the quality of its professionals, that is no small matter" [*ibid*, p. 59].

Examples of using manipulative materials presented in the following sections of this chapter demonstrate how the teaching of mathematics is inherently connected to the three descriptors of signature pedagogy—uncertainty, engagement, and formation [*ibid*, 2005]. The fact, that a mathematical problem may be too difficult to solve points at the uncertainty of mathematics pedagogy. By the same token, already one's attempt to solve the problem, even if only scanning through the diverse toolkit of problem-solving methods, is the engagement. Finally, through problem solving one forms a kind of a professional disposition towards mathematics, something that should stem from the very pedagogy of the subject matter. At the primary level (and beyond), problem solving and exploration can be assisted by manipulative materials the variety of which and a possible sophistication in their appropriate use, especially at the higher developmental levels, do bring uncertainty, engagement, and formation into the mathematics classroom.

3.5 Towards Rich Interpretations of Manipulative Representations

3.5.1 *Manipulative representation as text*

In the theory of sign-based (or semiotic) mediation [Wertsch, 1991], the word text refers to any verbal or nonverbal structure representing concrete situations through signs and symbols that mediate both memorization and comprehension. Within the elementary school mathematics curriculum, the wide range of tools including manipulative materials, drawings, diagrams, numbers, mathematical symbols and relations among them [National Council of Teachers of Mathematics, 1991] are the examples of text. That is, both the first-order and the second-order symbols are examples of text. All texts provide two basic functions, univocal and dialogic [Lotman, 1988]. The first function concerns the transfer of fixed meaning without expecting any inference to be drawn from it. The second function of text extends beyond the pure transferring of fixed meaning encouraging its multiple interpretations through which *new* meanings can be generated enabling one to go beyond memorization and, for example, "to begin developing an understanding of the commutative law as well as the mathematical practice of seeking structure" [Conference Board of the Mathematical Sciences, 2012, p. 9]. In the context of the elementary school mathematics, univocal function of text emphasizes "memorization of arithmetic "facts" [e.g., $3 + 1 = 4$ and $1 + 3 = 4$] with no attempt to encourage children to notice how [the] two facts might be related" [*ibid*, p. 9]. This connection, that is, an attempt to use dialogic function of text within which the commutative law is hidden, can be made through the first-order symbols by showing that regardless of the order in which one red counter and three yellow counters are counted, we have four counters.

Seeing manipulative representations as text, consider, once again, the diagram of Fig. 3.9 with two-color counters. A univocal perspective on text does not suggest any relation between the two lines of the counters. However, from a dialogic perspective, there are many ways to relate them. Their recognition requires certain level of mathematical competence on the part of a teacher, something that can (and must) be developed within a mathematics teacher education course. As the first-order symbols, both combinations of six counters represent a pattern in which colors alternate. Moreover, each combination yields another one through a simple flip of the counters. A transition to the second-order symbolism brings about many more interpretations. For example, in accord with a current focus on the development of algebraic thinking in the early grades mathematics through learning to "decompose numbers ... in more than one way ... by using objects or drawings" [Common Core State Standards, 2010, p. 11], each combination may be construed as a representation of the equality $6 = 1+1+1+1+1+1$. In that way, the de-contextualization of the fact that six counters are equal to one red plus one yellow plus one red plus one yellow plus one red plus one yellow can eventually lead to the discussion of different ways to decompose six (or a whole number n, in general) into a sum of non-negative integers.

Fig. 3.10. Revealing intuitively the hidden meaning of a manipulative task.
(Source: [Abramovich, 2010b, p. 54]).

For example, a second grader, who was asked to arrange four two-sided counters in all possible ways (two of which are shown in Fig. 3.9), not only found almost all arrangements (missing just one) but also attempted to extract new meaning from the text of his solution by using the word "brothers" (Fig. 3.10). Using this word, the second grader was able to reveal the hidden substance of the task, which was to create conditions for de-contextualizing (and, eventually, generalizing) by moving from the first-order symbols to the second-order symbolism. By interpreting each combination of counters numerically, one can develop understanding of what changes and what remains invariant when the number of counters involved becomes a variable quantity. Seeing the two arrangements of the counters as "brothers" in a second grade classroom would have been more likely, if both arrangements had been (or were helped to be) seen as *AB*-patterns at the kindergarten level.

3.5.2 *From 'brothers' to Pascal's triangle*

The notion of 'brother' used by the second grader as a tool through which combinations of counters were related to each other, is equivalent to the notion of 'opposite' introduced by a third grader when building all four-cube towers out of cubes in two colors [Maher and Martino, 1996]. Likewise, this construction is isomorphic to representing n as the sum of non-negative integers not greater than n. Indeed, the combinations labeled 'brothers' (like those labeled 'opposites') are numerically identical to each other and when put in pairs have a unique numeric representation of n as a sum of non-negative integers. The recognition of this fact makes it possible, in general, to move from 2^n towers (or 2^n arrangements of two-color counters) to the conclusion that there exist 2^{n-1} ways to represent n as a sum of non-negative integers with regard to their order (including the self-representation). Indeed, n can be partitioned in m summands where m changes from 1 to n. In the case of $n = 6$, partition into summands can be interpreted as placing marks in the five spaces between the counters (Fig. 3.11). No mark means one summand, one mark means two summands, two marks mean three summands, three marks mean four summands, four marks mean five summands, and five marks mean six summands. The number of ways to place marks is defined through the concept of combination (Chapter 2): there are $C_5^0, C_5^1, C_5^2, C_5^3, C_5^4, C_5^5$ ways to put, respectively, no mark, one mark, two marks, three marks, four marks, five marks in five spaces. For example, $C_5^1 = \dfrac{5!}{1! \cdot (5-1)!} = \dfrac{4! \cdot 5}{4!} = 5$ (Chapter 2, formula (2.1)) – the number of ways to select one space (to put a mark) out of five spaces available. By the rule of sum (Chapter 2), the number of ways to partition n into non-negative integers is equal to the sum $C_5^0 + C_5^1 + C_5^2 + C_5^3 + C_5^4 + C_5^5$ the summands of which are the entries of the sixth line in Pascal's triangle with the sum equal to 2^5. This issue will be discussed in Chapter 5, section 5.6.

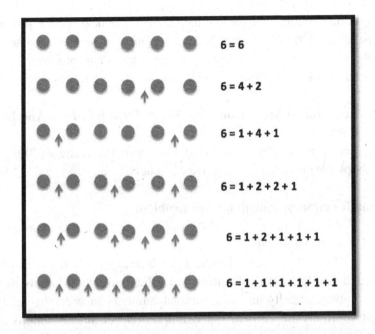

Fig. 3.11. Decontextualizing the placement of marks among six counters.

In that way, just as "the written language of children develops ... [by] shifting from drawings of things to drawings of words" [Vygotsky, 1978, p. 115], by using manipulative materials one is able to shift attention from pure diagrammatic representation (the first-order symbolism) of a problem situation to its numeric/algebraic representation (the second-order symbolism). In other words, the text of Fig. 3.11 can reveal, via the guidance of a 'more knowledgeable other', a "text within a text" [Lotman, 1988] structure. The most important issue associated with text, in general, and mathematical text, in particular, is how to make it function as a thinking device, generator of new meaning, and enabler of the development of new knowledge. A hidden connection between different layers of mathematical text becomes a clear-cut factor in the teacher's construction and the student's perception of a problem allowing for meaning-making processes to "take place as a consequence of an interaction between semiotically heterogeneous layers of a text" [*ibid*, p. 43]. In order for such text to start functioning as a thinking device for the student, the meaning of and connections among its different layers should first become apparent to the teacher. The next section provides an illustration of how revealing hidden meaning of text of a numeric task enables one to bridge conceptual understanding and procedural skills in the context of whole number arithmetic.

3.6 Learning to Move from One Type of Symbolism to Another and Back

The following problem is adopted from Van De Walle's [2001, p. 67] textbook for prospective elementary teachers.

A malfunctioning calculator key problem

Find the sum 354 + 541 + 435 by using a calculator with the malfunctioning 5-key.

Typically, teacher candidates see pedagogic text of this problem through purely numerical lenses and offer multiple solutions in which they enthusiastically add and subtract numbers in order to justify that their solutions are correct. For example, one of such solutions and its justification are as follows:

$$354 + 541 + 435 = 364 - 10 + 641 - 100 + 436 - 1$$
$$= (364 + 641 + 436) - 111 = 1441 - 111 = 1330.$$

How does one know that $354 + 542 + 435 = 1330$ without finding the sum in a purely numeric (operational) way? In fact, one cannot add the three numbers using a calculator because of the malfunctioning 5-key. So, teacher candidates verify the correctness of their solutions by adding and subtracting digit five free integers; yet their enthusiasm fades with each new response. Here, one can start experiencing a boring aspect of mathematics (arithmetic) that stems from the generally positive learning framework 'single question – multiple answers'. It is this boring aspect of mental (or even calculator-facilitated) computations that can serve as a motivation for the introduction of a new idea (concept).

To this end, one can compare the number of ones, tens, and hundreds in the three addends and in the four-digit number 1330. There is a zero in the place of ones in the number 1330 and $4 + 1 + 5 = 10$ – the sum of the digits representing ones in the three original numbers -- has zero ones also. There is a three in the place of tens in the number 1330 and $5 + 4 + 3 + 1 = 13$ – the sum of the digits representing tens in the three original numbers plus one ten made out of their ones – has three tens also. There is a three in the place of hundreds in 1330 and $3 + 5 + 4 + 1 = 13$ – the sum of the digits representing hundreds in the three original numbers plus one hundred made out of their tens – has also three hundreds and, in addition, the last sum has one thousand, just as the number 1330. Establishing identity between the conceptually expected sum of the three numbers and their formally computed sum was based on the concept of place value, something that the operation of addition is based on. Once we see a multitude of combinations of three digit numbers, the modification of which does not change the total number of ones, tens, hundreds, and thousands, the problem can be turned into a hands-on activity.

Unfortunately, the lack of experience with manipulative materials prevents teacher candidates from seeing this problem as an activity on a place-value chart that follows the rules of base-ten arithmetic. Such activity can be demonstrated in class as shown in Fig. 3.12. A simple, yet conceptually profound principle that governs the activity is that the counters may only be moved up and down but not left or right and none of the nine cells may include either five or more than nine counters. Once again, the teaching method here is based on the framework 'single question – multiple answers', but, in addition, the activity demonstrates how conceptual understanding facilitates procedural performance and turns the latter into a game. That is, the activity is first carried out conceptually at the level of the first-order symbols, and then each combination of counters is described procedurally through the second-order symbols.

The correct (up and down) movement of counters is easier to control and verify than to add three-digit numbers, especially when this has to be carried out more than one time. What one needs to check when using counters is that none of the cells has either five or more then nine counters (the former case can be recognized by subitizing [Kaufman *et al.*, 1949], a term describing an accurate visual quantitative judgment of a small number of objects without counting them), move them up and down only, and read and/or write each new combination of three 3-digit numbers having the same sum. In addition, the use of multicolored counters can be recommended for this activity to allow for each place value to be associated with a single color.

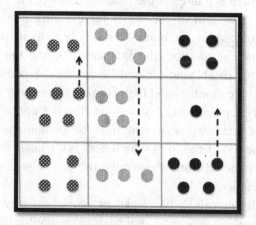

Fig. 3.12. Conceptual understanding in action:
$354 + 541 + 435 = 444 + 442 + 434$.

3.7　Perimeter and Area Using Square Tiles

The activity of this section is based on the use of square tiles in exploring the concepts of perimeter and area. Consider the combination of five square tiles forming a cross (Fig. 3.13). Assuming the side length of a tile equal to one linear unit, the task is as follows: *Augment the cross with like tiles so that a new shape has perimeter 16 linear units and all outer edges have the same (unit) length.*

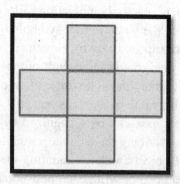

Fig. 3.13. Augmenting perimeter problem.

The first few things that have to be clarified are: What do we mean by perimeter? What is the perimeter of the cross? What do we mean by augmentation of the cross? Fig. 3.14 illustrates incorrect augmentations because not all outer edges of the so constructed shape measure the unit length. In fact, this is a problem that uses counting skills to answer a geometric question. In some sense, it is similar to explorations associated with the triangle inequality (section 3.2) where arithmetic skills were used in application to geometry.

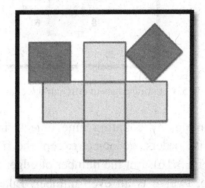

Fig. 3.14. Examples of incorrect augmentation.

The next step is to find the perimeter of the cross by counting edges. Although counting skills are mostly procedural ones, they do include some basic conceptual understanding that the order in which objects are counted is immaterial for the total count. In this regard, an interesting episode that occurred in a mathematics teacher education classroom through counting edges to find the perimeter of the cross is worth mentioning. Two students were counting edges in different directions by labeling them with numbers and someone in the classroom pointed out that an edge (different from the first one) received the same label, seven (Fig. 3.15). The discussion of whether counting objects in different directions always results in an object getting the same (numeric) label twice then ensued. The immediate question to be answered was as follows: If one augments the cross by other tiles (like it is shown in Fig. 3.19) and then counts edges in different directions starting from the same edge, would there always be an edge labeled by the same number? Once again, the diversity of ways to count edges motivated an inquiry into a new mathematical situation.

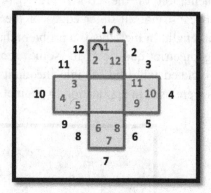

Fig. 3.15. Counting edges in different directions makes sevens meet.

For example, by counting nine objects in the clock-wise and counter clock-wise orders, no object except the first one gets the same numeric label (Fig. 3.16). But the number of edges in the cross is twelve. And unlike nine, twelve is an even number. Likewise, when counting lined-up objects in different directions (left-right and right-left), only in the case of an odd number of objects (Fig. 3.17) there is an object that gets the same label twice. In the case of counting an even number of lined-up objects, no object gets the same label twice (Fig. 3.18). The situation of counting on a clock is just opposite. In order to answer the question about whether the cross and its various augmentations always have an edge labeled by the same number (different from one) when counting in different directions requires an additional clarification that will be provided below.

To this end, a new way of counting perimeter of an augmented cross has to be discussed. Once it is established that the cross has perimeter 12 (linear units), a typical response to the task is to add tiles in different (correct) ways as shown in Fig. 3.19. Typically, teacher candidates find whether they have perimeter 16 by the direct counting of outer edges as shown in Fig. 3.15. Through this process (with the direction of counting being immaterial), different cases can be recorded in a chart (Fig. 3.20). However, after some time, their joy of counting edges fades and this gives the course instructor an opportunity to introduce a new way of counting; that is, counting tiles rather than edges.

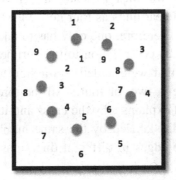

Fig. 3.16. Counting nine objects in different clock directions.

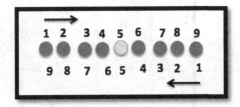

Fig. 3.17. Counting nine lined-up objects in different directions.

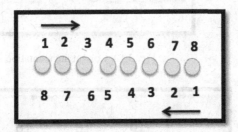

Fig. 3.18. Counting eight lined-up objects in different directions.

Note there are only three cases when augmentation by tiles does (or does not) change the total number of edges. The first case is when a pair of old and new tiles shares a single edge – this augments perimeter by two linear units. Indeed, in that case three new edges can be observed and one old edge disappears. The second case is when such a pair shares two edges – this does not change perimeter at all. Indeed, in that case, while two new edges can be observed, two old edges disappear. Finally, it is possible for two tiles to share a vertex only – this augments perimeter by four linear units as four new edges can be observed and all old edges remain. Therefore, one only has to take a look at an added tile in order to decide its contribution to the perimeter. In other words, one has to count tiles that were added and assign to them one of the three possible values: zero, two, or four. All the three numbers are even; in particular, this fact explains why the cross and its various augmentations always have an edge labeled by the same number (different from one) when counting the edges in different directions starting from the same edge.

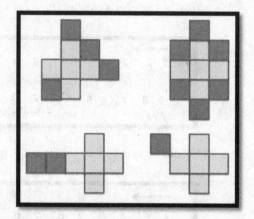

Fig. 3.19. Correct augmentation.

Perimeter	16	16	16	16
New tiles	4	6	2	1

Fig. 3.20. Chart-recording data.

Now the next set of questions can be asked: What is the smallest number of new tiles that yields perimeter 16 linear units? What is the largest number of tiles that yields this perimeter? Is there more than one way to place the largest number of tiles? Is it possible to add tiles to the cross in order to have a square with perimeter 16? If such square exists, how many tiles are needed to build it?

Obviously, the chart of Fig. 3.20 includes the smallest number of tiles, which is one. However, the largest number of tiles is not included in the chart. The construction of square with perimeter 16 is possible. Moreover, such a square will have 16 tiles. This square is remarkable because no other square built out of square tiles has area numerically equal to perimeter. In fact, as known from the history of mathematics, there are only two integer-sided rectangles with such a property. As recorded by Plutarch – a Greek historian of science of the 1st-2nd centuries AD – and cited in Van der Waerden [1961], "The Pythagoreans also have a horror for the number 17. For 17 lies halfway between 16 … and 18 … these two being the only two numbers representing areas for which the perimeter (of the rectangle) equals the area" (p. 96). Making historical connections in a mathematics teacher education course is not straightforward and as the above example with the cross demonstrates "the greater the mathematical sophistication of children [as well as teacher candidates], the more history is available to help consolidate their learning of mathematics" [Gardner, 1991, p. 17].

3.8   The Importance of Teacher Guidance in Using Manipulative Materials by Students

The importance of teacher as a 'more knowledgeable other' in helping students to use effectively the first-order symbols as the sine qua non of using the second-order symbolism was recognized by Clements [1999] who suggested that manipulative materials are pedagogically valuable when "used in the context of educational tasks to actively engage children's thinking with teacher guidance" (p. 56). Indeed, as was demonstrated through several examples in this chapter (as well as throughout the book), using these tools does help learners of mathematics to grow conceptually. However, for such growth to occur, teachers themselves have to be prepared to guide children in the right direction. The availability of manipulative materials in the mathematics classroom is a necessary but not sufficient condition for their appropriate use. As mentioned by Ball [1992], "creating effective vehicles for learning mathematics requires more than just a catalog of promising manipulatives" (p. 18). Similarly, Boggan *et al.* [2010] noted, "Children must understand the mathematical concept being taught rather than simply move the manipulatives around" (p. 3). Within a mathematics teacher education methods and content course, manipulative materials can also be introduced as support system in comprehending situations already expressed through the second-order symbolism. In doing so, multiple roles of the same type of manipulative can be demonstrated in illuminating the meaning of different concepts.

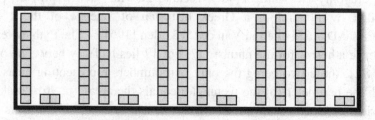

Fig. 3.21. The meaning of a long may depend on context.

As an example, consider the number sequences 2, 7, 12, 17, 22, 27, 32, 37, ... and 2, 7, 3, 8, 4, 9, 5, 10, ... (adopted from [Van De Walle *et al.*, 2010, p. 15]). One can see that the latter sequence consists of the sums of digits of the former sequence. Indeed, beginning from the third term in each sequence, we have $1 + 2 = 3$, $1 + 7 = 8$, $2 + 2 = 4$, and so on. Furthermore, every second number in the first and the second sequences increases, respectively, by ten and by one. These patterns can be explained by using base ten blocks (Fig. 3.21) – units and longs (ten units). However, in each case, a long would play a different role. For the sequence 2, 7, 12, 17, 22, 27, 32, 37, ...; skipping every second term can be represented through augmenting a number with a long, thus keeping the number of units (either two or seven) in the augmented number the same. So, Fig. 3.21 may represent four numbers – 12, 22, 32, 42 – that share the same last digit and differ from each other by a multiple of ten. When representing the sequence 2, 7, 12, 17, 22, 27, 32, 37, a long takes new meaning, a counter, when digits (face values) are added. Alternatively, Fig. 3.21 may also represent another four numbers – 3, 4, 5, 6 – demonstrating the meaning of adding digits. In the next section, the use of manipulatives in creating non-decimal representations of numbers as a way of conceptualizing the decimal system of arithmetic will be discussed.

3.9 Conceptualizing Base-Ten System Using Manipulative Materials

Hands-on experimentation with manipulative materials was found helpful in developing conceptual understanding of the decimal system of arithmetic through constructing diagrammatic representations of operations with whole numbers in a base system. Here, once again, one can turn to Vygotsky whose remarkable argument in favor of the in-depth teaching of arithmetic in primary schools is worthy presenting in full.

> "As long as the child operates with the decimal system without having become conscious of it as such, he has not mastered the system but is, on the contrary, bound by it. When he becomes able to view it as a particular instance of the wider concept of a scale of notation, he can operate deliberately with this or any other numerical system. The ability to shift at will from one system to another (e.g., to "translate" from the decimal system into one that is based on five) is the criterion of this new level of consciousness, since it indicates the existence of a general concept of a system of numeration. In this as in other instances of passing from one level of meaning to the next, the child does not have to restructure separately all of his earlier concepts, which indeed would be a Sisyphean labor. Once a new structure has been incorporated into his thinking— usually through concepts recently acquired in school—it gradually spreads to the older concepts as they are drawn into the intellectual operations of the higher type" [Vygotsky, 1962, p. 115].

Although Schmittau and Vagliardo [2006] have registered their concern in connection with the absence of the multi-base blocks from the modern classroom (base-ten blocks, shown in Fig. 3.21, are still widely available) something that, in their opinion, resulted from the influx of the ideas of the back to basics movement [Brodinsky, 1977], the appropriate use of square tiles can replace effectively these missing blocks. Moreover, square tiles are even more useful tools for comprehending the mechanism of transition from one base to another as they allow for a greater flexibility in the use of hands-on strategies.

As an illustration, consider the following task of creating the first-order symbols as support system in using the second-order symbolism of a base system.

In Fig. 3.22, a base-ten number 19 (one long of ten units plus nine units) is represented as a sum of four products (rectangles) of two numbers. Translate each product in base five and add (in that base) the resulting numbers. Represent each step both visually and symbolically.

One can see that in Fig. 3.22 there are two rectangles having more than four blocks; thus, their representation in base five should have two parts (digits), one of which is a long consisting of five units. This observation is reflected in Fig. 3.23, which translates rectangles (products) into base-five representations. A transition from one type of symbolism to another type allows for the following relations to be recorded: $13_5 = 2 \cdot 4$ and $11_5 = 2 \cdot 3$. These relations may be confusing for those used to be only operationally (rather than conceptually) proficient with the decimal system of arithmetic. Yet, the two relations are supported by the corresponding first-order symbols that follow the rules of the base-five system of arithmetic and on that ground they should be accepted as true multiplication facts for that base. Finally, transition from Fig. 3.23 to Fig. 3.24 completes the task implying the equality $19_{10} = 34_5$.

As one teacher candidate, reflecting on the course activities where the above ideas were introduced, noted: "*Manipulatives are a* physical representation of mathematics. The numbers, symbols, and names of functions that we give to these representations are simply an interpretation, a method of language, and a means of recording the physical representation. While these different bases seemed like a foreign language while completing this assignment, this is how young children feel as they begin to learn the symbols and representations of the base-ten numbers. Base-ten seems like the only way to represent numbers, because it is what we have always known, and it is all that we have ever known. Manipulatives, or the raw physical representation, is the only common ground we will have with young children as they begin to count, add, multiply, square numbers, or compute more complex activities". In her reflection, the teacher admits the primordial role of manipulatives which are relatively independent on the setting within which they are employed and represent "*the only common ground*" among different systems of arithmetic. Their "*raw physical*" nature, however, allows for multiple roles to be assigned. This, one more time, points at the importance of teacher guidance in students' use of manipulative materials.

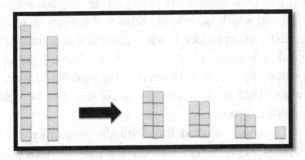

Fig. 3.22. Base ten: $19_{10} = (2 \cdot 4 + 2 \cdot 3 + 2 \cdot 2 + 1 \cdot 1)_{10}$.

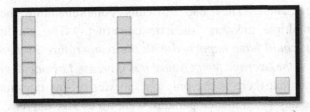

Fig. 3.23. Transition to base five: $(2 \cdot 4 + 2 \cdot 3 + 2 \cdot 2 + 1 \cdot 1)_{10} = 13_5 + 11_5 + 4 + 1$.

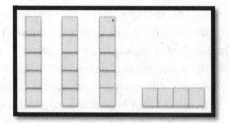

Fig. 3.24. The final result: $19_{10} = 34_5$.

3.10 Modeling as a Way of Creating Isomorphic Relationships

The use of manipulative materials in a mathematics classroom can also be interpreted in terms of modeling as one creates a physical representation (using the first-order symbols) of a mathematical situation to be explored and then interpreted using the second-order symbolism. Through this process, one creates isomorphic relationships not only between mathematics context and its physical model, but also between different representations of the context. After creating a physical model, one explores its properties, develops methods of investigation, comes up with a theory, and then applies the theory to the original context. There are several ways in which isomorphic relationships can be modified in a modeling context. One approach is when altering a situation yields a new model (physical or symbolic) and new methods of investigation. Another approach is when one begins with altering the model and its methods of investigation and then finds a new situation that can be described in terms of the modified model.

As an example of modeling with the first-order symbols, the following classroom episode is worth sharing [Abramovich *et al.*, 2012]. A group of second graders was working on a project dealing with the concept of average (arithmetic mean) as a characteristic of outdoor temperature change. They had difficulty comprehending the 'single question—multiple answers' didactic construct. The children were asked: *What could have happened with the temperature in the duration of five days if the average temperature has increased by one degree?* The response was that there is the only one answer to this question: There was just one day when the temperature went up by five degrees. Without a proper intervention, the second graders were unable to overcome the concreteness of a single day, not to mention their inability to grasp a possible split of temperature increase over several days.

Such intervention took place in the form of the following hands-on task: *Find all the ways to put five rings on the index and middle fingers*. Experimentally, using rings, the children found the answer and recorded it in the form of drawings (Fig. 3.25): five rings on the index finger; five rings on the middle finger; one ring on the index finger, four rings on the middle finger; one ring on the middle finger, four rings on the index finger; two rings on the index finger, three rings on the middle finger; two rings on the middle finger, three rings on the index finger. Furthermore, they were able to connect the rings activity with the possibility to accumulate the increase of temperature by five degrees over two days. With the appropriate teacher guidance, they constructed an isomorphic model using rings and fingers, recorded this model in terms of the first-order symbols, and then applied their findings to the temperature change context by writing all representations of the number 5 as a sum of two non-negative integers with regard to their order: $5 = 5 + 0, 5 = 0 + 5, 5 = 1 + 4, 5 = 4 + 1, 5 = 2 + 3, 5 = 3 + 2$.

Next, the children attempted to change the model by themselves, namely, to find all the ways to put five rings on three fingers and then apply their findings to the case of a three-day accumulation of the temperature increase. This, however, turned out to be not just beyond their abilities to keep records of action, but such change of model required them to replace physical action by the use of formal mathematics: finding all the ways of assigning three different categories (fingers) to five identical objects (rings). This, of course, was beyond their mathematical abilities. But the question was naturally raised by one of the children and, once again, it demonstrated the need for teachers to be able to answer such a question at least in terms of providing information. With this in mind, using the "Y/B" model (Chapter 2), one can form a 7-letter word YYYYYBB where five letters Y correspond to five rings and two letters B correspond to the number of boundaries one needs to separate three fingers. The number of permutations of the letters in this word is equal to $\dfrac{7!}{5! \cdot 2!} = \dfrac{5! \cdot 6 \cdot 7}{5! \cdot 2} = 21$ (formula (2.3), Chapter 2).

One child, using trial and error, found eight ways of accumulating five degrees over three days and, when she learned that there exist 21 ways, exclaimed "I won!" apparently indicating that among all the children trying to find all ways to accumulate the temperature increase by five degrees over three days, she was the closest one to the right answer. Such classroom experience gives meaning to the Conference Board of the Mathematical Sciences [2001] recommendation, "Prospective teachers ... should learn how basic mathematical ideas combine to form the framework on which specific mathematics lessons are built" (p. 8). That is, any new mathematical content has the potential to bring about new methods of teaching this content.

Fig. 3.25. Five rings on two fingers
(source [Abramovich *et al.*, 2012]).

Chapter 4

We Write What We See (W⁴S) Principle

4.1 Introduction

Mathematics, with its origin in the study of number and shape, has evolved from concrete activities to abstract concepts by means of argument and computation [Alexandrov, 1963]. The first mathematical problems, as known from history, stemmed from the contexts of counting using "the principle of one-for-one correspondence ... without a need for names for numbers" [Rudman, 2007, p. 31]. Later, the physical manipulation of objects and visual argumentation regarding the relationship among the objects led to the need for names describing specific properties of numbers. For example, "being ancient even in Plato's time [380 B. C.] ... [was the game of] guessing odd or even with respect to the number of coins or other objects held in hand" [Smith, 1953, p. 16] and a geometric term gnomon, resembling a sundial (an instrument that determines the time of the day by the position of the Sun), was used to refer to an odd number because of its double plus one form. Over the centuries, the development of mathematical knowledge evolved by taking into account the primordial nature of concrete objects including geometric shapes over the secondary nature of words and other signs that describe specific combinations and properties of those objects. Just as the teaching of writing was recommended to "be arranged by shifting the child's activity from drawing things to drawing speech" [Vygotsky, 1978, p. 115], the teaching of mathematics can be arranged as a transition from seeing and acting on concrete objects (the first-order symbols) to describing the visual and the physical through culturally accepted mathematical notation (the second-order symbols).

To this end, the "we write what we see" (W⁴S) principle can be proposed as a didactic motto to be used in the teaching of mathematics. We see differences and similarities when dealing with geometric figures or their images and appreciate different terminology to describe them; we see a relationship (known as the triangle inequality) among the lengths of three straws when trying to construct a triangle that does not (or does) allow for such a construction (Chapter 3, section 3.2); we see within a numeric table that the sum of two consecutive triangular numbers (Chapter 6, section 6.3.2) is the square of the rank of the larger number, an observation credited to a 4th century Greek mathematician Theon and used in the 18th century by a Dutch minister of church and mathematics teacher Élie de Joncourt to compute squares and square roots [Roegel, 2013]. There are plenty examples of that kind in school mathematics and beyond.

Yet, the W^4S principle works not without reservations. Although the avowal 'I see' often confirms understanding, mathematical visualization, as Tall [1991] put it, "has served us both well and badly" (p. 105). Therefore, while the focus on visualization is a commonly accepted practice of mathematics teaching and learning [e.g., Presmeg, 2006; Zimmermann and Cunningham, 1991], especially in the digital era, there is a long and sometimes challenging path from seeing things to understanding correctly their mathematical meaning or the absence thereof.

The advent of computers in the classroom has provided great many opportunities for visualizing mathematical concepts using software programs for the construction of graphs, diagrams, geometric shapes, numeric tables, and even step-by-step solutions to complex problems. Notwithstanding, the appropriate use of software is not a simple matter and a teacher has to possess both mathematical and technological skills in order to, whenever possible, provide students with conceptually accurate images of mathematical ideas under study. Likewise, a deep knowledge of mathematics is required to provide infallible visualization in off-computer environments. As Wittmann put it, "The most important thing in teaching is to understand mathematical structures as teaching aids that facilitate learning" (cited in [Akinwunmi *et al.*, 2014, pp. 361-362]). This chapter describes some pedagogical ideas grown along the above lines and, as in other parts of the book, born in the context of the author's work with prospective teachers of mathematics.

4.2 W⁴S Principle and the Duality of Its Affordances

The limitations of uncritically using the W^4S principle become obvious already at the pre-school level. Indeed, we can see that two pineapples are bigger than three plums. As one moves from visual to symbolic, the first and the second kinds of fruit can be associated with the numbers 2 and 3, respectively. But this does not imply that labels attached to the numbers may be dropped leading one to conclude that in the domain of the second-order symbolism two is bigger (greater) than three. This is a simple example of how the adage 'a picture is worth a thousand words' may be misleading in the absence of conceptual understanding of the dual nature of educational affordances, positive and negative, that a picture provides.

In what follows, the duality of affordances of the W^4S principle in the teaching of mathematics will be discussed. The theory of affordances [Gibson, 1977] is frequently used nowadays when talking about teaching mathematics with computers [Kieran and Drijvers, 2006; Lingefjärd, 2012; Watson and Fitzallen, 2016]. However, what is true for a computer environment is also true for any learning environment. In general, the more positive affordances an educational environment offers, the fewer negative affordances it presents. At the same time, negative affordances of a particular pedagogical approach are often hidden and an uncritical use of any approach can lead one astray in the learning of mathematics, thereby increasing the effect of hidden didactic challenges. In order to minimize negative affordances of a learning environment, the ability to conceptualize the first-order symbols created through action is crucial.

Kadijevich and Haapsalo [2000] referred to a case when procedure is informed by concept as an educational approach to the teaching of mathematics. Such conceptually informed procedure may include the creation of the first-order symbols toward the end of developing their interpretation through the second-order symbolism. For example, an algorithm of finding the number of ways a doctor can schedule eye exams for three people (alternatively, the number of permutations of three different objects) can be informed by the rule of product (Chapter 2), which, in turn, can be presented in the form of a tree diagram – a visual demonstration of using the rule in developing permutations. Conceptual understanding plays the critical role not only in seeing things in terms of understanding them but creating educationally flawless visual representations of mathematical concepts. By the same token, as will be shown in this chapter in the context of developing procedures for multiplying and dividing fractions, conceptual perspective on visual representations can turn negative affordances of a learning environment into its positive affordances. In Chapter 3 (section 3.6) such was an example of a calculator with a malfunctioning key when the negative affordances of seeing the problem through purely computational lenses were eliminated through an insight offering the positive affordances of the combination of a place value chart with multicolored counters and the W^4S principle.

Mathematical knowledge develops from action on concrete objects to their formal description through words and/or mathematical notation. So, in the teaching of mathematics one may encourage students at all levels to start doing mathematics from acting on the first-order symbols and then, through the appropriate use of the W^4S principle, make a transition to the second-order symbolism abstracted from the concreteness of visual representations. Technology, both digital and physical, plays an important role in this transition and therefore, it has to be used appropriately. To this end, the appropriate use of technology can be defined as balancing positive and negative affordances of what technology (which may include more than one digital or non-digital tool) provides; ideally, maximizing positive affordances and minimizing negative affordances of the tools. In the case of computers, because it is a teacher who "has a critical responsibility in shaping the relation between the computational media and mathematical knowledge" [Guin and Trouche, 2002, p. 200], courses for prospective teachers of mathematics must provide guidance on how to shape this relation starting from the very basic examples of using the W^4S principle. The same is true regarding the relation between non-digital teaching aids and mathematical concepts they are designed to support.

4.3 W^4S Principle in Teaching Primary School Mathematics
As was mentioned above, even in rather simple situations, the W^4S principle might give misleading results in the absence of conceptual

understanding. In fact, conceptual understanding can be fostered through the use of counter-examples (see Chapter 9 for more details): a combination of two pineapples and three plums can be used to develop the appreciation of the concept of unit in modeling the relationship between (or among) whole numbers. Quantitatively, those objects (pineapples and plums) are not comparable due to different units they represent. Likewise, without recourse to the notion of experimental probability (Chapter 8) one cannot easily respond to the question about chances of randomly picking a plum from a basket with three plums and two pineapples. That is, seeing a picture does not mean that, in the absence of conceptual understanding, one can describe it quantitatively in a correct way.

In order to develop such understanding of comparing quantities, a teacher can give students linking cubes (non-digital technology) of the same size to construct towers 2-cube tall and 3-cube tall and then ask the following questions:

- What do you see? Which tower is taller?
- What can be said about the numbers 2 and 3? Which one is bigger?

In the digital era, comparing quantities can be facilitated (and conceptually enhanced) by using a computer program, such as *The Geometer's Sketchpad* (created by Nicholas Jackiw in the late 1980s and still commonly used in the schools in North America) which can help one to construct towers out of the same size squares. Here, the main idea is to have students construct squares all the same size by appropriately using the construction features of the program. Note that the W^{4}S principle works the same way for the fruit and the squares when the total number of objects has to be determined: in both contexts one can see without any reservation that $2 + 3 = 5$. Conceptual difficulties with addition begin with the introduction of a base system that can be overcome through the appropriate use of the W^{4}S principle in the context of base-ten (or multi-base) blocks [Abramovich, 2012].

The Geometer's Sketchpad can also be used to write a simple program for comparing fractions through fraction circles integrating conceptual understanding into a computer-mediated action. Consider the task of constructing the fraction circles 1/2, 1/4, 1/6, and 1/12 (by defining the location of their center and the length of the radius), arranging them from the least to the greatest, and finding their sum. Both operations should be presented in iconic and numeric forms.

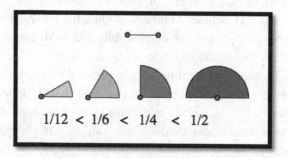

Fig. 4.1. We write what we see.

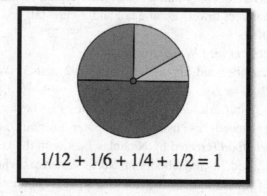

Fig. 4.2. Addition automatically requires equal radii.

The main focus of this task deals with a frequently overlooked (or taken for granted) fact that fractions may be compared only when the same unit is their point of reference. In terms of icons (the first-order symbols), the fraction circles 1/2, 1/4, 1/6, and 1/12 are four different sectors cut off from the same whole circle. In terms of the second-order symbols, these four (unit) fractions are parts of the same unit. This idea is hidden in the construction of a fraction circle when one defines its radius. As shown in Fig. 4.1, the four fraction circles have the same radius pictured at the top of the sketch. Therefore, they may be compared both as the first-order symbols (icons) and the second-order symbols (numbers). Finally, when adding the four fractions (Fig. 4.2), their radii can be adjusted through an action that brings about a new picture, the description of which in the domain of the second-order symbols is then made in terms of a numeric equation that becomes independent of any context. Here, the concept of same unit is implicitly embedded in the construction of the very unit using its parts. That is, unlike comparing fractions, adding them has to be carried out correctly already in the domain of the first-order symbols. Indeed, as shown in Fig. 4.3, when adding fraction circles representing different units, one is apparently unable to describe the sum at the level of the second-order symbolism. Furthermore, unlike five fruits result from adding two pineapples and three plums, no meaningful numeric combination results from adding half and third of, respectively, a pineapple and a plum. This is another example of the unity of content and methods: teaching to add (or subtract) fractions as parts of the same unit requires methods different from teaching to add (or subtract) whole numbers.

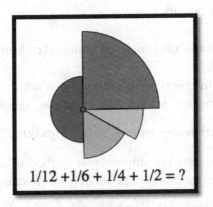

$$1/12 + 1/6 + 1/4 + 1/2 = ?$$

Fig. 4.3. The sum does not make sense.

The above use of fraction circles emphasizes the fact that when one carries out arithmetical operations with fractions, it is assumed, though tacitly, that they are fractional parts of the same unit. In that way, using the computer program not only "require[s] the user to *describe* intended relationships" [Goldenberg and Cuoco, 1998, p. 365, italics in the original] but also forces one to integrate meaning with the required construction of the objects of visualization. Therefore, a conceptual flaw occurring at the action level (constructing fraction circles with different radii) might result in the erroneous ordering of the objects based on their size, followed by an incorrect symbolic description through an uncritical use of the W^4S principle. In the case of adding such fraction circles (Fig. 4.3), no symbolic description of the sum can be offered. So, only the equal radii construction of fraction circles allows for their comparison, otherwise the comparison of visual images is meaningless. At the same time, the operation of adding fraction circles representing different units as the first-order symbols does not yield *any* result at all at the level of the second-order symbolism. It is not surprising that, conceptually, equalities are considered being more sophisticated entities of mathematics than inequalities. Indeed, a child learns first that $\dfrac{1}{2} > \dfrac{1}{3}$ because understanding of this inequality is based on perception; much later, the child learns that $\dfrac{1}{2} + \dfrac{1}{3} = \dfrac{5}{6}$ because understanding of this equality is based on operation.

4.4 Comparing Non-Unit Fractions Using Area Model

4.4.1 *Comparing fractions being close to each other*

As a more complicated arithmetical example, consider the case of comparing the fractions $\dfrac{3}{5}$ and $\dfrac{4}{7}$ using a picture. Which fraction is bigger? Fig. 4.4 shows a one-dimensional method of the comparison of fractions using the so-called area model for fractions. Whereas this method works well for comparing basic unit fractions such as $\dfrac{1}{2}$ and $\dfrac{1}{3}$, for non-unit fractions that are sufficiently close to each other, this method stops working because of its dependency on the accuracy of drawing. As shown in Fig. 4.4, by looking at the representation of the fractions, $\dfrac{3}{5}$ and $\dfrac{4}{7}$, from left to right one sees what can be described symbolically as $\dfrac{3}{5}<\dfrac{4}{7}$; by looking at the same representation of the

fractions from right to left one sees the opposite relation, $\dfrac{3}{5}>\dfrac{4}{7}$.

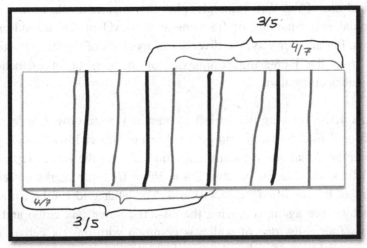

Fig. 4.4. Visualization depending on accuracy in drawing is contradictory.

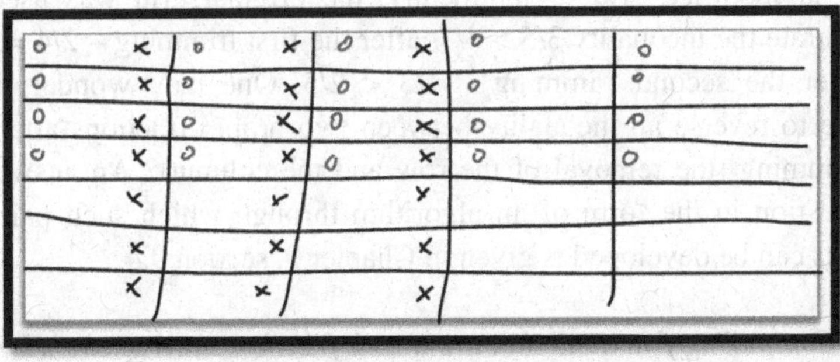

Fig. 4.5. Comparison of fractions through counting marks.

Rather than discouraging the use of the W⁴S principle, this example shows the deficiency of the one-dimensional representation of the comparison of fractions and thereby, it motivates and justifies the need for a two-dimensional method shown in Fig. 4.5. So, negative affordances of the one-dimensional representation were used as a counter-example in comparing fractions and motivated the introduction of a two-dimensional representation using area model and the appreciation of the positive affordances of the method. That is, the misleading diagram of Fig. 4.4, serving as a counter-example to the uncritical use of the W^4S principle, plays an important role in fostering conceptual understanding of fractions at the level of the second-order symbols. Put another way, the diversity of methods of teaching fractions stems from the increasing complexity of the content of elementary mathematics curriculum.

Remark 4.1. Removing the far-left column and the top row (sharing the top-left cell) of the 35-cell grid pictured in Fig. 4.5 yields a 24-cell grid in which the x-marks represent the fraction 2/4 and the zeroes represent the fraction 3/6. That is, because 2/4 = 3/6 = 1/2, both marks represent one-half of the modified (trimmed) grid. The latter grid can be trimmed further by, once again, removing the far-left column (six cells) and the top row (four cells, one of which is common with the six cells). This trimming yields a 15-cell grid in which the remaining x-marks and zeroes represent 1/3 and 2/5 (alternatively, 5/15 and 6/15), respectively, of the so modified grid. Symbolically, the original grid was used to demonstrate the inequality 3/5 > 4/7, after the first trimming – 2/4 = 3/6, and after the second trimming – 1/3 < 2/5. One may wonder: is it possible to reverse an inequality between two proper fractions after the first trimming (the removal of the row and the column)? An answer to this question in the form of an algorithm through which such pairs of fractions can be developed is given in Chapter 9, section 9.4.

4.4.2 *Comparing fractions that are a unit fraction short of the whole*
There are cases when fractions can be compared indirectly by comparing them through benchmark fractions. Unit fractions are commonly used as benchmark fractions [Common Core State Standards, 2010]. Which fraction is bigger: $\dfrac{4}{5}$ or $\dfrac{5}{6}$? A special property of the two fractions is that each is a unit fraction short of the whole. That is, $\dfrac{4}{5} + \dfrac{1}{5} = 1$ and

$\dfrac{5}{6}+\dfrac{1}{6}=1$. Because $\dfrac{1}{5}>\dfrac{1}{6}$, the fraction $\dfrac{5}{6}$ is closer to the whole than $\dfrac{4}{5}$.

Therefore, $\dfrac{5}{6}>\dfrac{4}{5}$.

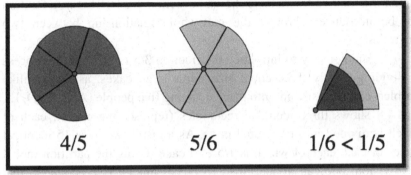

Fig. 4.6. Comparing fractions which are a unit fraction short of the whole.

4.5 From Comparison of Fractions to Arithmetical Operations Using Area Model

4.5.1 *Fractions as part-whole and divisor-dividend models*

There are two ways through which the concept of a proper fraction can be introduced. The first way is based on a part-whole relationship where, in a real-life context, a fraction measures, for example, the chances of a random outcome. Indeed, we commonly say that there are three chances out of five to pick up (without looking) a red M&M from a bag with five M&Ms, three of which are red and two are yellow. Whereas it is clear without any fractions that the chances for a red candy are higher than for a yellow candy (because three is greater than two; although, this does not imply that one would always pick up a red candy), one needs to somehow measure chances when they have to be compared to the case of another bag with, say, seven M&Ms four of which are red. In this simple situation with M&Ms, it is for the purpose of understanding which outcome is more likely, that chances for different bags have to be measured by numbers – quantities which, in turn, can be compared. To this end, as shown in Fig. 4.7, the bag with five M&Ms can be represented as a chart divided into five equal parts three of which are filled with red M&Ms and the remaining part with, say, two yellow candies. That is, the part of the chart filled with red M&Ms can be represented (measured) by the fraction 3/5. Likewise, the chances for the second bag can be measured by the fraction 4/7. (The method of comparing the two fractions is shown in section 4.4.1). That is, a fraction can be introduced through the part-whole relationship between two quantities.

Another way to introduce the fraction 3/5 is through the process of dividing evenly three equal size items (e.g., cakes, as they, unlike marbles, can be easily cut into pieces) among five people (A, B, C, D, E). Fig. 4.8 shows three identical rectangles (representing cakes), each of which is divided into five equal parts. As a result, we have 15 identical pieces of cake each of which is 1/5 of a cake. Using the partition model for division, one can divide those 15 pieces among five people so that each person would get three pieces. In order to answer the question what fraction of a cake would each person get, one has to keep in mind that (as was already mentioned) one piece is 1/5 of a cake (the whole that gives birth to the unit fraction 1/5, the latter being a part of the former) and therefore, each person would get 3/5 of a cake. That is, in the second case the fraction 3/5 describes the result of dividing three by five; in other words, $3 \div 5 = \dfrac{3}{5}$. Put another way, the fraction 3/5 (or, in general, m/n, $m < n$) emerges from the process of dividing 5 into 3 (or, in general, n into m). In particular, when such a division has a zero remainder, the corresponding fraction turns into an integer. This connection between fraction and division is mentioned by the Conference Board of the Mathematical Sciences [2012, pp. 25, 28] as one of the fundamental ideas that teachers teaching grades 3 – 5 need to know well.

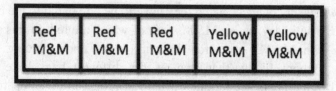

Fig. 4.7. The chances to pick up a red candy are measured by 3/5.

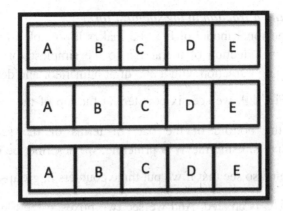

Fig. 4.8. Each person gets 3/5 of a cake.

4.5.2 *The concept of common denominator*

One can recognize in the diagram of Fig. 4.5 the concepts of the common denominator of two fractions (a cell of the grid is a unit of measurement of the fractions) and the product of the fractions (the overlap of two different marks), something that is definitely missing in Fig. 4.4. Such recognition of the basic concepts of arithmetic is due to the power of the W^4S principle, which works well within a flawless learning environment of the first-order symbols.

To clarify the common denominator aspect of the diagram of Fig. 4.5, note that any cell of the grid is one out of 35 ($= 5 \cdot 7$) cells that comprise the grid; that is, as a fraction, a cell is equal to $\frac{1}{35}$ assuming that the whole grid is the unity. It is a cell that serves as a common measure of the grid's parts, which we called $\frac{3}{5}$ and $\frac{4}{7}$ in Fig. 4.5, something that the one-dimensional diagram of Fig. 4.4 does not offer. Therefore, in Fig. 4.5, the region marked with x's takes $\frac{21}{35}$ of the grid and the area marked with zeroes takes $\frac{20}{35}$ of the grid. Put another way, $\frac{21}{35} = \frac{3}{5}$ and $\frac{20}{35} = \frac{4}{7}$. That is how the concept of the common denominator can be introduced through the first-order symbols.

4.5.3 *Reducing a fraction to the simplest form*

The two-dimensional model of Fig. 4.5 makes it possible to explain the meaning of the reduction of a fraction to the simplest form. We will explain this using a fraction with a one-digit numerator and denominator, say, $\frac{6}{9}$. In Fig. 4.9, we see six counters at the top of the bar and nine counters at the bottom of the bar. In terms of the second-order symbolism, that is, using the W^4S principle, we describe the diagram as $\frac{6}{9}$. But we can also see that if we put three counters in groups of three, a new unit, a box, is created. And we see two boxes at the top of the bar and three boxes at the bottom of the bar. What is important here is that each box comprises the same number of counters. At the same time, the fraction $\frac{6}{7}$ is not reducible further because seven counters may not be put into groups of two or three, something that can be done with six counters. The same is true for the fraction $\frac{4}{9}$ although neither numerator nor denominator is a prime number. If two sets of objects cannot be put in groups with the same number of objects in each group, then at the level of the second-order symbolism the two sets are described by relatively prime numbers and, therefore, the corresponding fraction is already presented in the simplest (irreducible) form.

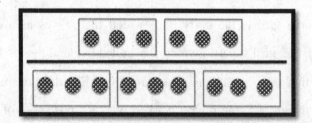

Fig. 4.9. Reducing fraction to the simplest form as a change of unit.

4.5.4 *Using unit fractions as benchmark fractions*

The use of unit fractions as benchmark fractions involves several ideas. The first idea is estimation. When the exact value of a quantity in relation to other quantities (the numeric value of which is easier to comprehend) is not known (although numeric value of a quantity is unique), then different ways of comparison can be used. To begin, consider the case of adding whole numbers. How to estimate the sum 16 + 27 using consecutive multiples of ten as benchmark numbers between which the sum belongs? Because 16 has only one unit of ten and 27 has only two units of ten, we know that $16 < 20$ and $27 < 30$ from where it follows that their sum has fewer than five units of ten, that is, $16 + 27 < 50$. By the same token, $16 > 10$ and $27 > 20$ from where it follows that their sum is greater than three units of ten, that is, $16 + 27 > 30$. That is, the sum can be estimated as follows: $30 < 16 + 27 < 50$. However, 30 and 50 are not consecutive multiples of ten and thus one needs to replace either 30 or 50 by 40. To this end, one can recall that a possible representation of 10 as a sum of two integers is $10 = 5 + 5$ and therefore one can compare 16 and 27 to the nearest multiples of five. The inequalities $16 > 15$ and $27 > 25$ imply the inequality $16 + 27 > 15 + 25 = 40$. Thus the sum $16 + 27$ satisfies the inequalities $40 < 16 + 27 < 50$ where 40 and 50 as consecutive multiples of ten are used as benchmark numbers in estimating the "unknown" sum.

Returning back to fractions, note that an obvious estimation of a fraction by a unit fraction is not always straightforward. Instead, a fraction may be estimated by another fraction, which is equal to the unit fraction after the reduction to the simplest form. This is similar to the case of estimating the sum $16 + 27$ using multiples of five rather than multiples of ten and then making the latter from the former; that is, moving from five being a unit of comparison to ten being such a unit. Because the reduction of a fraction to the simplest form means the change of unit, the second idea behind using unit fractions as benchmark fractions is the change of unit.

Consider the case of estimating the fraction 3/7. On the one hand, $3/7 < 3/6$ – we made the fraction 3/7 bigger by making its denominator smaller. On the other hand, we have $\dfrac{3}{6} = \dfrac{3 \cdot 1}{3 \cdot 2} = \dfrac{1}{2}$ – a unit fraction where both numerator and denominator consist of three old units. That is, $3/7 < 1/2$. Likewise, $3/7 > 3/9$ – we made the fraction 3/7 smaller by making its denominator bigger. By the same token, we have

$\dfrac{3}{9} = \dfrac{3 \cdot 1}{3 \cdot 3} = \dfrac{1}{3}$ – a unit fraction where both numerator and denominator

consist of three old units. That is, $\dfrac{1}{3} < \dfrac{3}{7} < \dfrac{1}{2}$ where 1/3 and 1/2 are

benchmark (consecutive unit) fractions.

The use of unit fractions as benchmark fractions can be demonstrated visually by using a computer program enabling the construction of fraction circles. Fig. 4.10 shows how 3/7 can be placed in the space between two consecutive unit fractions, 1/2 and 1/3. Similarly (Fig. 4.11), using technology, the estimate 1/5 < 5/24 < 1/4 can be first obtained through trial and error and then formally articulated using the language of fractions: the smaller (greater) – the greater (smaller). That is, 5/24 > 5/25 = 1/5 and 5/24 < 6/24 = 1/4.

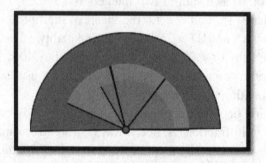

Fig. 4.10. Benchmark fractions for 3/7 are 1/2 and 1/3.

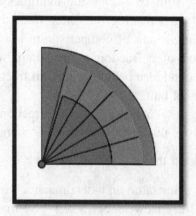

Fig. 4.11. Benchmark fractions for 5/24 are 1/4 and 1/5.

4.5.5 *Multiplying fractions using area model*

Another conceptually important aspect of the two-dimensional diagram of Fig. 4.5 is that it offers visual means for understanding the rule of multiplying fractions. The goal here is to develop a rule indicating that the product of two fractions is a fraction the numerator and denominator of which are, respectively, the products of numerators and denominators of the factors. For example, how can one see the product $\frac{3}{5} \cdot \frac{4}{7}$ in the diagram of Fig. 4.5? Why is this product, a fraction, represented by the region where both marks overlap? The answer is in the meaning of multiplication that does not change as a number system changes; that is, a physical meaning of multiplication is the same for objects described by integers as by fractions. What is $3 \cdot 4$ at the action level? It is taking three groups of four objects. That is, the multiplication sign at the action level is characterized by the preposition 'of'; that is, taking one quantity *of* another quantity. Likewise, $\frac{3}{5} \cdot \frac{4}{7}$ means taking $\frac{3}{5}$ of $\frac{4}{7}$; that is, taking the quantity 3/5 *of* the quantity 4/7. To make this operation more concrete, in other words, by demonstrating the skill of contextualization of the operation, one can first take $\frac{4}{7}$ of a cake (the whole) and then take $\frac{3}{5}$ of the piece taken. This action results in a new piece of the cake characterized by the product $\frac{3}{5} \cdot \frac{4}{7}$. This is a piece where both marks, zeroes and x's, overlap after 3/5 is marked with x's and 4/7 – with zeroes. Now we see the region described by the product, but how can one describe the region through a single fraction? The overlap of x's and zeroes can be seen as 3 groups of 4 objects, that is, $3 \cdot 4 = 12$ cells belong to the overlap. The total number of cells can be seen as 5 groups of 7 objects, that is, $5 \cdot 7 = 35$ cells comprise the whole grid. Therefore, by demonstrating the skill of de-contextualization, the overlap as a fraction of the grid (the whole) can be expressed numerically through the fraction $\frac{12}{35}$. In other words, $\frac{3}{5} \cdot \frac{4}{7} = \frac{3 \cdot 4}{5 \cdot 7} = \frac{12}{35}$. That is, the rule (algorithm) of multiplying fractions at the level of the second-order symbolism was developed conceptually by using the first-order symbols allowing one to see "where a mathematical rule comes from" [Common Core State Standards, 2010, p. 4].

Consider another example when one has to multiply two improper fractions using the area model. How can one see the product $\frac{5}{3} \cdot \frac{7}{4}$ using a grid? As before, we start with drawing a rectangle (the borders of which are solid lines) to represent the whole (Fig. 4.12). The next step is to divide the rectangle (vertically) into four equal parts, each of which is 1/4 and then extend it to the right by another three-fourths to get 7/4 (marked with zeroes). Then, the so constructed fraction 7/4 is divided (horizontally) into three equal parts each of which is 1/3 of 7/4 and then extend it down by another two-thirds to get the product $\frac{5}{3} \cdot \frac{7}{4}$.

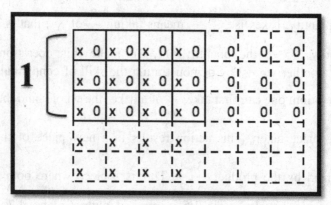

Fig. 4.12. The product $\frac{5}{3} \cdot \frac{7}{4} = 1 + \frac{3}{4} + \frac{2}{3} + \frac{3}{4} \cdot \frac{2}{3}$.

One can see that the cells marked with x's represent 5/3. The cells marked with only x's and with only zeroes represent, respectively, 2/3 and 3/4. The cells with no marks represent the product $\frac{2}{3} \cdot \frac{3}{4}$, which, by using 6 as a new unit, can be reduced to the simplest form 1/2. Put another way, the diagram of Fig. 4.12 reflects the distributive property of multiplication over addition:

$$\frac{5}{3}\cdot\frac{7}{4}=(1+\frac{2}{3})(1+\frac{3}{4})=1+\frac{2}{3}+\frac{3}{4}+\frac{2}{3}\cdot\frac{3}{4}.$$

Finally, one can see that the overlap of 5/3 and 7/4, that is, the part of the diagram of Fig. 4.12 with both marks coincides with the whole. Unlike the case of the product of two proper fractions when their overlap was their product located within the whole, the overlap of the product of two improper fractions, as shown in Fig. 4.13, is the whole itself.

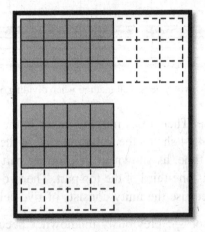

Fig. 4.13. The overlap of 7/4 and 5/3 is the whole.

4.6 Dividing Fractions Using Area Model

4.6.1 *Partition model for division supports contextualization*

Let us begin with dividing whole numbers using area model. What does it mean to divide 5 into 3? Because 5 > 3, the smaller number cannot be measured by the larger number as one of the interpretations of division suggests. That is, in terms of the measurement model for division, the inclusion of 5 into 3 is an abstraction, as something larger may not be included in its part. Nonetheless, this abstraction can be represented in the form of a (positive) fraction smaller than one. Contextualization may help to comprehend abstraction. In our case, the partition model for division means dividing three things among five people. Obviously, the things should be divisible, in general; that is, conducive to be partitioned in smaller parts. For example, it is not possible to divide three marbles among five people. Yet, it is possible to divide three (identical) cups of cottage cheese among five people. To do that, one has to divide each cup into five equal parts, giving each person 1/5 of a cup. With three cups, each person would get 3/5 of a cup. Put another way, in order to find how many times 5 is included into 3 one has to divide 3 by 5.

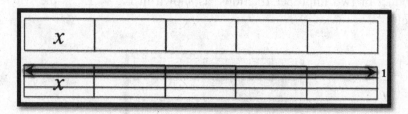

Fig. 4.14. Using two-dimensional method when dividing 5 into 3.

Let $3 \div 5 = x$. Then x is a missing factor in the equation $5x = 3$. The top part of Fig. 4.14 shows the left-hand side of the last relation, that is, $5x$. At the same time, as shown in the bottom part of Fig. 4.14, the unity (shaded dark) is one-third of the top part. The two-dimensional area model shows that because the unity consists of five cells, each of which represents the number $\frac{1}{5}$, previously unknown x becomes known as it consists of three such cells. That is, $x = \frac{3}{5}$ or $3 \div 5 = \frac{3}{5}$.

Now, using the same method, let us divide two proper fractions. For example, let us find the value of $\frac{2}{3} \div \frac{4}{5}$. The result of this division is a number x, four-fifth of which is equal to two thirds, that is, x is the missing factor in the equation $\frac{4}{5}x = \frac{2}{3}$. Fig. 4.15 comprises three diagrams, the far-left one representing x. The middle diagram shows that x consists of five equal sections with an unknown numerical value. In order to make it known, one has to represent the unity. Because, $\frac{4}{5}x$ (shown in the middle diagram) is equal to $\frac{2}{3}$ (of the unity), it has to be

extended by $\dfrac{1}{3}$. This extension is shown in the far-right diagram of Fig.
4.15 which, once again, uses two-dimensional method. Through this
method, the unity turned out having been divided into $4 \cdot 3 = 12$ equal
sections. At the same time, we see that x comprises $5 \cdot 2 = 10$ such
sections. That is, $x = \dfrac{2}{3} \div \dfrac{4}{5} = \dfrac{2 \cdot 5}{3 \cdot 4} = \dfrac{10}{12}$. Put another way, in order to
divide $\dfrac{2}{3}$ into $\dfrac{4}{5}$ one has to multiply $\dfrac{2}{3}$ by the reciprocal of $\dfrac{4}{5}$. In
section 6.3 below, the meaning of this rule, commonly referred to as
invert and multiply, will be explained.

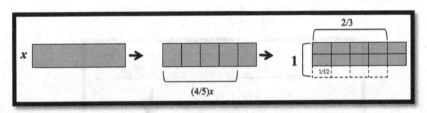

Fig. 4.15. Using two-dimensional method when dividing $\dfrac{2}{3}$ by $\dfrac{4}{5}$.

Remark 4.2. In the case of multiplying proper fractions using the two-
dimensional area model, the picture multiplication begins with drawing
the unity. In the case of dividing proper fractions, the unity is not known
and it has to be found through solving an equation with a missing factor.
The same is true for improper fractions (see Fig. 4.19 below).

4.6.2 *The importance of unit in solving word problems with fractions*
When solving word problems involving fractions using the first-order
symbols, once again, one has to critically rely on the W^{4}S principle.

Problem 1. Jerry has four bottles of apple juice. If the serving
glass is 3/5 of a bottle, how many servings can he make out of the
bottles?

Solution. As shown in Fig. 4.16, in order to solve the problem,
one has to measure four bottles by a glass, that is, to use the
measurement model for division when dividing 3/5 into 4. This division
yields the remainder. An uncritical use of the W^{4}S principle may

describe the remainder (two shaded sections within the fourth bottle) as 2/5. However, one cannot describe servings both in terms of glass and bottle. As shown in the far-right diagram of Fig. 4.16, the shaded piece is 2/3 of a glass (serving). That is, from four bottles of apple juice Jerry can make $6\frac{2}{3}$ servings.

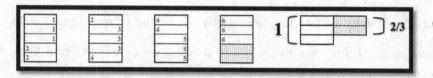

Fig. 4.16. Six and two-thirds glasses is the serving.

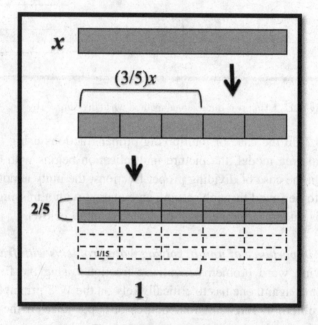

Fig. 4.17. Dividing 2/5 by 3/5.

Alternatively, as a practice in dividing fractions using the two-dimensional model, one can measure 2/5 by 3/5; in other words, one has to divide 2/5 by 3/5, that is, to find x from the equation

$$\frac{3}{5}x = \frac{2}{5}. \tag{4.1}$$

The process of solving Eq. (4.1) is shown in Fig. 4.17 where the top diagram represents x, the middle diagram shows the left-hand side of the equation, and the bottom diagram shows that the equality between the two sides of Eq. (4.1) implies the following: the unity consists of 15 sections and therefore x consists of 10 sections each of which is equal to $1/15$. As a result, we have $x = \frac{10}{15} = \frac{2}{3}$. Note that a formal, purely arithmetical solution to Problem 1 does not result in any confusion. Indeed, $4 \div \frac{3}{5} = 4 \cdot \frac{5}{3} = \frac{20}{3} = 6\frac{2}{3}$. (Invert and multiply rule has yet to be explained).

Problem 2. Mary has $6\frac{2}{3}$ pounds of flour. It takes $2\frac{1}{2}$ pounds to make a cake. How many cakes can she make?

Solution. As shown in Fig. 4.18 using the measurement model for division, Mary can make two full cakes. After that, $1\frac{2}{3}$ pounds of flour remain. Once again, the question is how to measure this amount by the full cake, that is, how to measure a fraction by a fraction. Simply put, what is the result of dividing $2\frac{1}{2}$ into $1\frac{2}{3}$? To answer this question, one has to solve the equation $2\frac{1}{2} \cdot x = 1\frac{2}{3}$ or

$$\frac{5}{2} \cdot x = \frac{5}{3}. \tag{4.2}$$

The process of solving Eq. (4.2), which is equivalent to finding $x = \frac{5}{3} \div \frac{5}{2}$, is shown in Fig. 4.19. The top diagram at the right shows the value of $\frac{5}{2} \cdot x$, with x shown at the top left. The bottom diagram shows that the equality $\frac{5}{2} \cdot x = \frac{5}{3}$ implies that the unity consists of 15 sections and therefore x consists of 10 sections each of which is $1/15$. Therefore, we have $x = \dfrac{10}{15} = \dfrac{2}{3}$, that is, Mary can make $2\dfrac{2}{3}$ cakes.

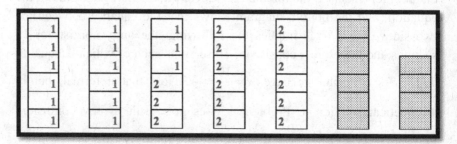

Fig. 4.18. What fraction of the cake can be made out of the shaded flour?

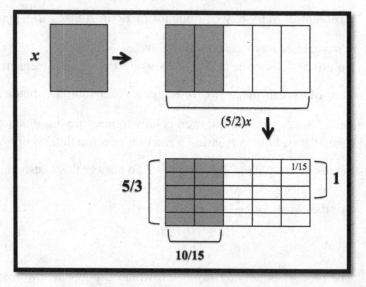

Fig. 4.19. Dividing $1\dfrac{2}{3}$ by $2\dfrac{1}{2}$.

4.6.3 *The meaning of "invert and multiply" rule*

Measurement model for division demonstrated in the diagram of Fig. 4.18 was applied to dividing $6\dfrac{2}{3}$ by $\dfrac{5}{2}$. Arithmetically, this division uses *invert and multiply* rule, $6\dfrac{2}{3} \div \dfrac{5}{2} = \dfrac{20}{3} \cdot \dfrac{2}{5} = \dfrac{8}{3} = 2\dfrac{2}{3}$, the meaning of which is the change of unit. To explain, let us consider the simpler problem to which the rule will be applied:

> *It takes two tiles to make a tower. How many towers can one make out of six tiles?*

An answer to this question results from the division, $6 \div 2 = 3$. Here, a tile is the unit and, therefore, a tower can be assigned the number 2. Measuring six units by two units yields three towers (Fig. 4.20). Now, let us do the division as follows: $6 \div 2 = 6 \cdot \dfrac{1}{2} = 3$. The meaning of replacing 2 by 1/2 can be explained in the following terms: we have six tiles each of which is one-half of tower; that is, out of six one-half towers three towers can be constructed. It is replacing tile by tower to serve as the unit (Fig. 4.20) that division is replaced by multiplication.

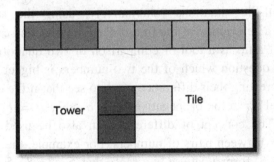

Fig. 4.20. Replacing tile as unit by one-half of tower.

The same rule and its meaning should work for fractions as well. The symbolic expression $6\dfrac{2}{3} \div \dfrac{5}{2}$ implies that the unit consists of two halves as shown at the top part of the diagram of Fig. 4.21. When we replace the above division by multiplication $6\dfrac{2}{3} \cdot \dfrac{2}{5}$, we assume that now the unit consists of five fifths and, therefore, in the context of Problem 2, instead of pound being the unit, we have assigned the role of unit to cake, thus making pound equal numerically to 2/5 instead of 1 (as shown in the bottom part of the diagram of Fig. 4.21).

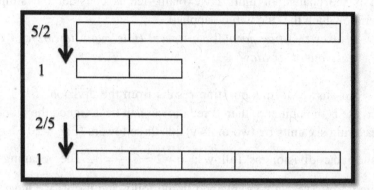

Fig. 4.21. Invert and multiply rule is a change of the whole.

4.7 Ratio and Proportion

Following the notion of introducing mathematical concepts as tools used to model real-life situations (alternatively, using the second-order symbolism as a description of the first-order symbols), one can introduce the concept of ratio as a tool of comparison of two quantities. This is similar to the question which of the two numbers is bigger that can be answered by creating their difference $a - b$ to see that if the difference is positive, equal to zero, or negative then $a > b$, $a = b$, or $a < b$, respectively. The concept of difference can also be used to describe commonalities between pairs of numbers. For example, the pairs (5, 2), (8, 5) and (11, 8) form a pattern in a sense that $5 - 2 = 8 - 5 = 11 - 8 = 3$, that is, the number 3 is the common difference for the three pairs.

Consider another three pairs of integers: (6, 2), (9, 3) and (12, 4). What do the pairs have in common? They are not in the same difference because $6 - 2 \neq 9 - 3 \neq 12 - 4$. Nonetheless, the three pairs do form a pattern because $6 \div 2 = 9 \div 3 = 12 \div 4 = 3$. This time, the number 3 shows how many times the larger number contains the smaller number. Such quantitative characteristic of the relation between two numbers is called ratio. Alternatively, the last chain of equalities can be represented in the following form

$$2 \div 6 = 3 \div 9 = 4 \div 12 = 1/3.$$

That is, the number 1/3 shows what fraction of the larger number is the smaller number. The diagram of Fig. 4.22 shows that the last chain of equalities is due to the change of unit: $2/6 = 1/3$ because two is a new one, $3/9 = 1/3$ because three is a new one, and finally, $4/12 = 1/3$ because four is a new one.

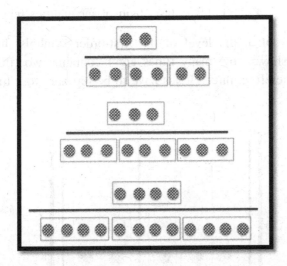

Fig. 4.22. Three pairs of different number of objects are in the same ratio.

Often, in a problem-solving situation, between two quantities one is unknown but the ratio of the quantities is known. Usually, the ratio is given in the simplest form, that is, as a fraction with numerator and denominator being relatively prime numbers. In this case, one can construct an equation called a proportion by equating the ratio of two quantities to the given ratio. For example, the diagram of Fig. 4.23 shows a pictorial approach to solving the problem: If the ratio of circles to squares is three to two, find the number of circles when the number of squares is eight. To find the number of circles, one begins with the smallest number of circles and squares in the given ratio; that is, one begins with three circles and two squares (shaded in Fig. 4.23). But there are eight squares, that is, four times as many as we have shaded, and in order to keep the ratio as given, we have to increase the number of (shaded) circles four-fold to get twelve circles. By making four a new one, the diagram shows that the ratio between twelve and eight is, indeed, three to two.

At the level of the second-order symbolism, the proportion from which the unknown quantity x can be found has the form $\dfrac{x}{8} = \dfrac{3}{2}$ whence $x = \dfrac{8 \cdot 3}{2} = \dfrac{8}{2} \cdot 3 = 4 \cdot 3 = 12$. The last chain of inequalities repeats all the steps carried out at the level of the first-order symbols. Indeed, the fraction 8/2 shows how many times eight contains two (that is, four times) and, therefore, three circles have to be repeated four times to get twelve.

Fig. 4.23. A typical ratio problem with an unknown quantity.

4.8 Percent and Decimal as Alternative Representations of a Fraction

The concept of percent can be introduced through an alternative way of comparing fractions the denominators of which are proper factors of 100, that is, one of the numbers 2, 4, 5, 10, 20, 25, 50. These factors can be paired so that the relations $2 \cdot 50 = 4 \cdot 25 = 5 \cdot 20 = 10 \cdot 10$ hold true. For example, to compare the fractions 3/4 and 7/10 one can use two 100-cell grids as shown in Fig. 4.24: the grid on the left shows the first fraction as 75/100 and the second fraction as 70/100. That is, 3/4 of the grid takes 75 cells and 7/10 of the grid takes 70 cells. The difference is five cells, which comprise 1/20 of the grid. Therefore, 3/4 is greater than 7/10 by

1/20. Note that here the unity is 100 and, therefore, what was one before, now becomes 1/100. In order to demonstrate this, for a fraction $p/100$ one uses the notation $p\%$ (reads p percent, from Latin *per centum* – by the hundred).

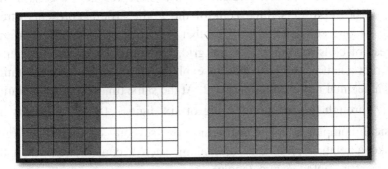

Fig. 4.24. Comparing the fractions 3/4 and 7/10 using 100-cell grids.

From here, one can make a transition to another representation of numbers called the decimal representation. A decimal representation of a fraction (or, more generally, a rational number) may have either finite or infinite number of digits. For example, a decimal representation of 1/2 is .5 (alternatively, .4999…, indicating a possibility of non-uniqueness of decimal representation for some rational numbers); the decimal representation of 1/3 is .333… (or .$\overline{3}$, which is not a different decimal representation but rather a short cut for repeating 3 using an ellipsis); the decimal representation for 1/7 is .142857142857… (or, .$\overline{142857}$), where the string of six different digits repeats itself over and over.

To explain the meaning of the notation .333… note that it is similar to the representation of a base-ten integer as a polynomial in the powers of 10 (e.g., $123 = 1 \cdot 10^2 + 2 \cdot 10^1 + 3 \cdot 10^0$) so that

$$.333... = 3 \cdot 10^{-1} + 3 \cdot 10^{-2} + 3 \cdot 10^{-3} + ... = \frac{3}{10} + \frac{3}{100} + \frac{3}{1000} + ... \,.$$

The last expression represents the sum of the geometric series with the first term 3/10 and the ratio 1/10. A closed form for this sum can be computed by dividing the first term by the difference between the unity and the ratio: $\dfrac{3/10}{1-1/10} = \dfrac{3}{9} = \dfrac{1}{3}$. This explains the equality 1/3 = .333… , where an empty space to the left of the decimal point means that the corresponding number is smaller than one.

Visually, the infinite sum $\dfrac{3}{10}+\dfrac{3}{100}+\dfrac{3}{1000}+...$ can be represented on a 100-cell grid in which 3/10 represents 30 cells of the grid, 3/100 represents three cells of the grid, 3/1000 represents three-tenths of one cell, and so on; in other words, it represents the infinite process of constructing 1/3 of the grid. Obviously, it is much easier to construct a half of the grid and this explains the shortness of the notation .5 as a decimal representation of 1/2. At the same time, the representation of 1/7 through the repeating string of six digits, $.\overline{142857}$, reflects the physical complexity of constructing one-seventh of a 100-cell grid (unlike a relative simplicity of making such a construction when the unity is represented as a rectangle).

One may ask as to why we need such complicated decimal notations instead of just using fractional notations. A possible answer is that the latter notation represents an operation and the former (decimal) notation represents the result of this operation in the decimal system of arithmetic. So, both the relations $5 + 6 = 11$ and $1/3 = .333...$ represent the result of binary operations (addition and division, respectively) on integers in a decimal notation system.

To conclude this section, an explanation of two decimal notations for 1/2 can be provided. Because

$$.0999... = \frac{9}{10^2}+\frac{9}{10^3}+\frac{9}{10^4}+... = \frac{9/10^2}{1-1/10} = \frac{9}{90} = \frac{1}{10}, \text{ we have}$$

$$.4999... = .4+.0999... = \frac{4}{10}+\frac{1}{10} = \frac{5}{10} = .5. \text{ That is, } \frac{1}{2} = .5 = .4999... .$$

4.9 Multiplying and Dividing Decimal Fractions

Using the two-dimensional area model for fractions on a 100-cell grid, one can visualize the meaning of placing the decimal point when multiplying decimal fractions. For example, a formal explanation of the equality $0.4 \times 0.2 = 0.08$ is either through counting the number of digits to the right of the decimal point of the factors or by translating decimal fractions into common fractions and back as follows:

$$0.4 \times 0.2 = \frac{4}{10} \times \frac{2}{10} = \frac{8}{100} = 0.08 \,.$$

The last chain of equalities can be represented on a 100-cell grid as shown in Fig. 4.25 in which the overlap of the pictorial representations of the factors occupy eight cells of the grid which symbolic representation is 0.08. It is not uncommon seeing students writing the result as 0.8. The W^4S principle enables a diagrammatic support of the comprehension of a rule described at the level of the second-order symbolism. One has only to represent correctly decimal fractions on a 100-cell grid and then count the number of cells that their overlap comprises. For example, the overlap of the diagrammatic representations of the fractions 0.4 and 0.8 comprises 32 cells, so the product of the two fractions is the fraction $\frac{32}{100}$ and in the decimal representation the decimal point gets its place immediately before the digit 3. But this place is due to what one sees rather than what one remembers through a formal rule. Of course, for more complicated decimal fractions their picture multiplication has a difficult representation in terms of the W^4S principle, but still, conceptual explanation can be provided. For example, consider the product 0.16×0.17. By analogy with the representation of the product 0.4×0.2 on a 100-cell grid where the number of cells is a product of denominators of the fractions $\frac{4}{10}$ and $\frac{2}{10}$, the product 0.16×0.17 can be represented on a 10^4-cell grid where the number of cells is the product of denominators of the fractions $\frac{16}{100}$ and $\frac{17}{100}$. An image of this product using the two-dimensional area model for multiplying fractions is shown in Fig. 4.26. The product is the sum $100 + 60 + 70 + 42 = 272$ – a number comprising $\frac{272}{10,000} = 0.0272$ of the 10^4-cell grid.

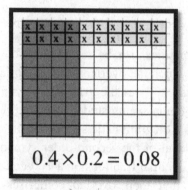

Fig. 4.25. Multiplying decimals using area model.

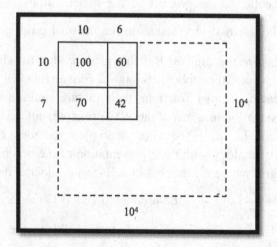

Fig. 4.26. Multiplying 0.16×0.17 on a 10^4-cell grid.

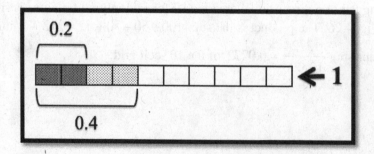

Fig. 4.27. Dividing decimal fractions using measurement model.

Consider now the case of dividing two decimal fractions. To this end, one can use either the measurement model for division or solving an equation with a missing factor. In the former case, when looking for the result of the operation $0.4 \div 0.2$ one can draw both fractions in the one-dimensional format to see how many times 0.2 is included into 0.4. Fig. 4.27 clearly shows that the result is 2. Likewise, in the latter case one can denote $x = 0.4 \div 0.2$ and then find x from the equation $0.2x = 0.4$ (Fig. 4.28). The diagram of Fig. 4.28 has to be explained. As before, x is an unknown quantity represented as a rectangle divided vertically into ten equal pieces two of which represent $0.2x$. At the same time, the last quantity is 4/10 of the unity. Thus 4/10 has to be augmented by 6/10 to get the unity. One can see that the unity comprises 20 equal cells. With each cell being 1/20 of the unity, we have $x = 40 \cdot \dfrac{1}{20} = 2$.

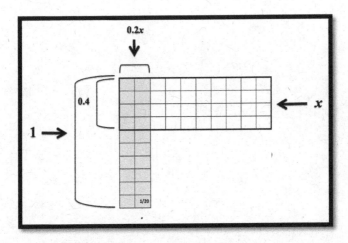

Fig. 4.28. Finding x from the equation $0.2x = 0.4$.

Chapter 5

Partitioning Integers into Like Summands

5.1 Introduction

One of the major ideas of the entire school mathematics curriculum is the representation of numbers as a combination of other numbers. Such representations serve different purposes, being important for mathematics itself and/or for real-life applications of mathematics. For example, representing an integer as a sum of two equal integers indicates that the integer is an even number. Less obvious is the fact that representing an integer as the sum of the first n odd integers indicates that the integer can serve as area of a $n \times n$ square. In the specific case of the fraction 1/2, its representation as a sum of three unit fractions indicates that their denominators can serve as the dimensions of a rectangular prism.

This chapter deals with partitions of positive integers into a number of like summands. Euler was the first to recognize the importance of this concept for the development of mathematics [Dunham, 1999]. In the modern classroom, using either physical or virtual manipulative materials, problems dealing with additive partitions can be introduced even before the study of arithmetic, providing young learners of mathematics with understanding the multiplicity of answers in the tasks that involve counting, informal learning to reason recursively in resolving quantitative queries, experience in recognizing different conditions under which partitioning problems can be formulated, and appreciating the innate complexity of the problems and their natural extensions already within a seemingly mundane context.

The material of this chapter goes beyond the pre-operational mathematics classroom with its only developmentally appropriate focus on hands-on explorations of quantitative situations that can be later described through abstract symbolic representations in other parts of the vast K-12 mathematics curricula. Following the viewpoint, "What students can learn at any particular grade level depends upon what they have learned before" [Common Core State Standards, 2010, p. 5], the chapter shows how the appropriate use of traditional teaching methods of elementary mathematics content can motivate the bottom to top approach to mathematics curricula through the study of various advanced concepts, including mathematical proof, difference equations, Pascal's triangle, binomial coefficients, and combinatorial counting techniques. By the same token, the top to bottom approach to mathematics curricula enables certain elements of higher concepts to be introduced at a lower level. With this in mind, the material of this chapter will show how one can use the concept of additive partition of integers to demonstrate a possible learning trajectory spanning from visual and hands-on (experiential) activities to symbolic to computational ones and then back to symbolic and/or experiential activities, yet at a higher cognitive level.

The chapter begins with a few problems (tasks) the hands-on solution of which requires just common sense and basic counting skills. Yet, one's possession of common sense does not imply the ability to reason systematically even when dealing with the first-order symbols, let alone the second-order symbols. That is why hands-on solutions to the first two tasks will be justified through formal reasoning and, in doing so, diversity in reasoning techniques will be demonstrated. This kind of transition from visual to symbolic through the use of the W^4S principle illustrates the power of mathematical concepts in verifying the results of a hands-on experiment. Even if those concepts are beyond one's immediate mastery, the appreciation of the power of conceptualization, abstraction and generalization when justifying experimental results enhances one's mathematical experience. Just as "mathematics courses [for prospective teachers] that explore elementary school mathematics in depth can be genuinely college-level intellectual experiences, which can be interesting for instructors to teach and for teachers to take" [Conference Board of the Mathematical Sciences, 2012, p. 31], the elementary school mathematics curriculum is a perfect place to "construct arguments ... [which] can make sense and be correct ... even though they are not generalized or made formal until later grades" [Common Core State Standards, 2010, p. 7].

5.2 Partition of Integers into Summands

A partition (alternatively, decomposition) of a positive integer n is a representation of n as a sum of positive integers. For example, the number 10 has many such partitions (e.g., $10 = 1 + 9$, $10 = 9 + 1$, $10 = 5 + 5$, $10 = 1 + 2 + 3 + 4$, $10 = 1 + 3 + 6$) each of which can be described conceptually in its own way. In all, there are 512 ($= 2^{10-1}$) different decompositions of 10 with regard to the order of summands, including the self-decomposition (Chapter 3, section 3.5.2). That is, as Ahlgren and Ono [2001] noted, "the seemingly simple business of counting the ways to break a number into parts leads quickly to some difficult and beautiful problems" (p. 1). Through the language of mathematics, different conditions under which such numeric representations are created can be formulated. Consequently, as has already been demonstrated in the book, different mathematical concepts may be taught under the umbrella of integer partitions. So, the relations $10 = 1 + 9$ and $10 = 9 + 1$ may be described in terms of commutativity of addition, the relation $10 = 5 + 5$ may focus, among other ideas, on the partition in two equal summands, the relation $10 = 1 + 2 + 3 + 4$ demonstrates a possibility of decomposition in consecutive integers (Chapter 6), the relation $10 = 1 + 3 + 6$ decomposes 10 in consecutive triangular numbers with the sum being a triangular number itself[8]. The above five integer partitions of 10 can first be represented through the first-order symbols as shown in Fig. 5.1.

In Fig. 5.2 both parts represent same thing: $10 = 1 + 3 + 6$. However, it is unlikely that an average student would see in the left-hand part something else than the sum of the three numbers. At the same time,

---

[8] An immediate question that one may ask is: are there other examples of that kind? For example, every second integer can be decomposed into equal summands. This fact can be easily explained: the number 10 is even and all even numbers possess this property. But is there another triangular number that can be represented as a sum of three consecutive triangular numbers? The trial and error approach yields another such example: $136 = 36 + 45 + 55$. However, the explanation of how this can be done in a systematic way is not easy. Computationally, other representations of a triangular number as a sum three consecutive triangular numbers can be found by using *Wolfram Alpha* (www.wolframalpha.com) – a computational knowledge engine available free on-line – through the query "solve over the integers $3n^2+9n+8=m^2+m$, $n>0$, $m>0$" where the quadratics in n and m are, respectively, twice the sum of three triangular numbers of ranks n, $n + 1$, $n + 2$, and that of rank m. Interesting, it is not possible to represent a square of a natural number as a sum of squares of three consecutive natural numbers.

its right-hand part shows what is special about the summands. In other words, the right-hand part is a representation within a representation. As mentioned by Uttal *et al.* [1997], "the difficulty that children encounter when using manipulatives stems from the need to interpret the manipulative as a representation of something else" (p. 38). In this regard, it is the teacher's responsibility to motivate a student to see this "something else". As a result, a student should be able to appreciate that the right-hand part is designed to demonstrate more than decomposition of an integer into the sum of three integers, and thereby, there is something special about the summands.

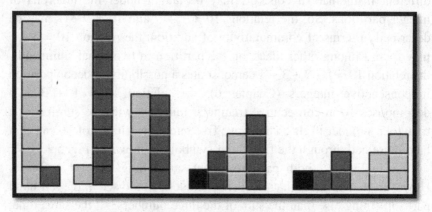

Fig. 5.1. Different ways to partition 10 into summands.

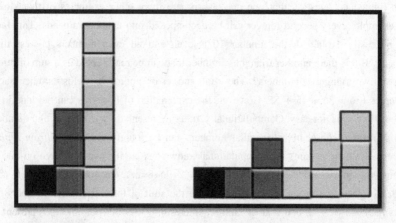

Fig. 5.2. Two ways to express a quantity, 10, through the first-order symbols.

Consider a few tasks formulated in the context of constructing towers out of square tiles – a typical standards-based setting recommended for mathematical practice in a junior elementary classroom aimed "to uncover students' thinking as they work with concrete materials by asking questions" [National Council of Teachers of Mathematics, 2000, p. 80], to help students "construct arguments using concrete referents such as objects, drawings, diagrams, and actions" [Common Core State Standards, 2010, p. 7], and to enable students' "proficiencies in mathematical skills … taught with an understanding of the underlying mathematical principles and not merely as procedures" [Ministry of Education, Singapore, 2012, p. 15]. Likewise, elementary teacher candidates enrolled in a mathematics methods and content course should come to appreciate that a variety of "tools exist to support teaching and learning … [and] to understand the mathematical aspects of these tools" [Conference Board of the Mathematical Sciences, 2012, p. 2].

Task 1. *Find all ways to build towers out of four square tiles and then arrange them in different orders.*

Solution. First, we have to explain what we mean by a tower made out of square tiles. It is a structure in which the tiles are lined-up both horizontally and vertically (e.g., see Fig. 3.6 in Chapter 3) so that the number of tiles at any vertical level in not smaller than in the next (higher) level. In other words, tiles may not be hanging above the ground level. Fig. 5.3 shows that there are eight ways to build towers out of four square tiles arranged in different orders.

Second, we have to explain that there are indeed eight towers that satisfy the condition of the task; that is, to show how one can construct towers through a system. To this end note, that there exist only one tower which is one-tile tall and only one tower which is four-tile tall. There exist four towers that are two-tile tall. Finally, there are two towers that are three-tile tall. This systematic reasoning is based on putting towers in groups according to their height.

The towers can also be put in groups according to their width. There exist only one tower four-tile wide and only one tower one-tile wide. There exist three towers that are three-tile wide. Finally, there exist three towers that are two-tile wide.

The third step is to describe the towers of Fig. 5.3 numerically using the W^{4}S principle. We have

$$4 = 1+1+1+1, \, 4 = 1+1+2, \, 4 = 1+2+1,$$
$$4 = 2+1+1, \, 4 = 2+2, \, 4 = 1+3, \, 4 = 3+1, \, 4 = 4 \, .$$

In section 5.6 of this chapter, another (formal) method of obtaining the above eight representations of four as a sum of non-negative positive integers will be presented. However, while considering four as a special case, the method would be able to essentially give the total number of such representations, rather than the representations themselves. Indeed, the method would allow one to find this number not through the development of an organized list (either through building towers or by adding numbers) which becomes a non-method for a large number of tiles but by using a formula that relates a positive integer n to the number of representations of n through a sum of non-negative integers.

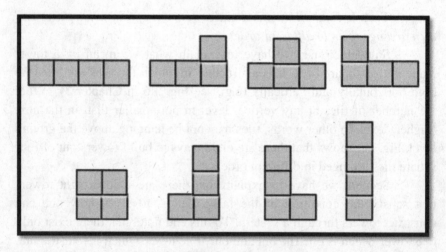

Fig. 5.3. Solution to Task 1.

Task 2. *Find all ways to build towers out of four square tiles arranged in the non-decreasing order.*

Solution. One way to solve this task is to remove from the towers constructed under the condition of Task 1 all the towers that do not satisfy the condition of Task 2. The result is shown in Fig. 5.4 where the extraneous towers are shaded dark. This solution is of particular interest for the future explorations. It shows that using four tiles, one can build (under the condition of Task 2) four towers in one way, three towers in one way, two towers in two ways, and one tower in one way. Another method to solve Task 2 is to build the five towers and then prove that all towers were constructed by focusing, say, on their height. To this end, note there is only one tower that is four-tile high; there is also only one tower that is three-tile high; there are two towers that are two-tile high; finally, there is only one tower that is one-tile high. A similar reasoning can be used to arrange towers according to their width. Using the W^4S principle, the (not shaded dark) towers in the diagram of Fig. 5.4 can be described as follows:

$$4 = 1 + 1 + 1 + 1, 4 = 1 + 1 + 2, 4 = 2 + 2, 4 = 1 + 3, 4 = 4.$$

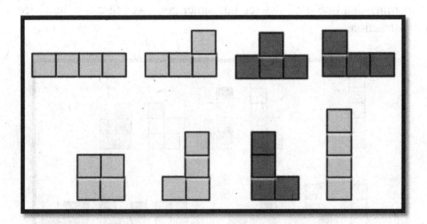

Fig. 5.4. Solution to Task 2: extraneous towers shaded dark.

In all, there are five ways to build towers out of four square tiles and arrange them in the non-decreasing order. It should be noted that partitioning of an integer into a sum of non-negative integers without regard to order (Task 2), in comparison to partitioning with regard to order (Task 1), turns out to be a more difficult concept. Towards the end of developing the former concept, consider

Task 3. *Find all ways to build towers out of five square tiles and then select only those towers in which the tiles are arranged in the non-decreasing order.*

Solution. The diagram of Fig. 5.5 shows that out of five tiles one can construct 16 different towers among which there are only seven towers with tiles arranged in the non-decreasing order (those extraneous are shaded dark). Numerically, using the W⁴S principle, these seven towers can be represented as the following partitions of five into all summands without regard to order

$$5 = 5, 5 = 1+4, 5 = 2+3, 5 = 1+1+3, 5 = 1+2+2,$$
$$5 = 1+1+1+2, 5 = 1+1+1+1.$$

The total number of towers, 16, shown in Fig. 5.5 will be justified formally in section 5.6 below. In a mean time, note that the towers were still constructed through a system (an organized list) focusing on either their width or height. A formal justification, however, will not be based on an organized list to allow for dealing with any number of square tiles, something that one does not see but rather perceives through the process of abstraction.

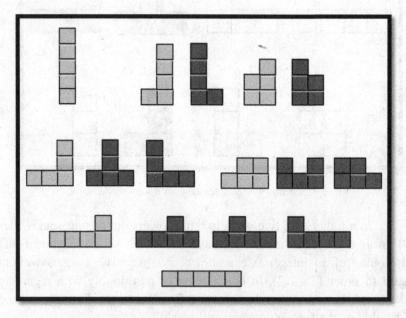

Fig. 5.5. Solution to Task 3: extraneous towers shaded dark.

Task 4. *Find all ways to build three towers out of eleven square tiles and arrange then in the non-decreasing order.*

Solution. Fig. 5.6 shows that three towers can be constructed out of eleven tiles in ten ways. The towers are constructed through a system which is reflected in the following numeric description of Fig. 5.6:

$$11 = 1 + 1 + 9 = 1 + 2 + 8 = 1 + 3 + 7 = 1 + 4 + 6 = 1 + 5 + 5 = 2 + 2 + 7$$
$$= 2 + 3 + 6 = 2 + 4 + 5 = 3 + 3 + 5 = 3 + 4 + 4,$$

where the first number, 1, is combined with all representations of ten (11 − 1 = 10) as a sum of two positive integers (the upper part of Fig. 5.6); then the number 2, used first, is combined with all representations of nine (11 − 2 = 9) as a sum of two positive integers excluding the unit (the first three towers of the bottom part of Fig. 5.6); then (finally) the number 3, used first, is combined with all representations of eight (11 − 3 = 8) as a sum of two positive integers excluding integers smaller than three (the last two towers).

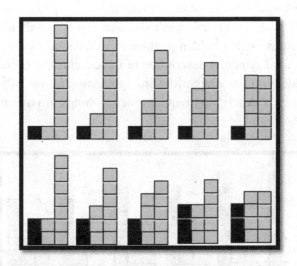

Fig. 5.6. Partition of eleven into three parts according to the smallest addend.

One can generate these ten partitions of eleven by using another strategy (Fig. 5.7). There is only one tower 9-tile high because the remaining two tiles (11 − 9 = 2) can be decomposed in two towers in a single way: 2 = 1 + 1. There is only one tower 8-tile high because the remaining three tiles (11 − 8 = 3) can be decomposed in two towers in a single way as well: 3 = 1 + 2. There are two towers 7-tile high because the remaining four tiles (11 − 7 = 4) can be decomposed in two towers in two ways: 4 = 1 + 3 and 4 = 2 + 2. There are two towers 6-tile high because the remaining five tiles (11 − 6 = 5) can be decomposed in two towers in two ways as well: 5 = 1 + 4 and 5 = 2 + 3. There are three towers 5-tile high because the remaining six tiles (11 − 5 = 6) can be decomposed in two towers in three ways: 6 = 1 + 5, 6 = 2 + 4, and 6 = 3 + 3. Although seven tiles (11 − 4 = 7) can be decomposed in two towers in three ways, the only possible decomposition to go with a tower 4-tile high is 7 = 3 + 4. Once again, ten different towers have been constructed. Fostering different types of systematic reasoning supports the development of "two complimentary abilities to bear on problems involving quantitative relationships: the ability to *decontextualize* – to abstract a given situation and represent it symbolically … and the ability to *contextualize* … in order to probe into the referents for the symbols involved" [Common Core State Standards, 2010, p. 6, italics in the original]. The diversity of methods and tools used to solve a manipulative task with a hidden mathematical meaning opens a window to a more general symbolic description of the conclusions of the task that is not dependent on the straightforward application of the W^4S principle but rather, such a description requires seeing things through the lens of abstraction.

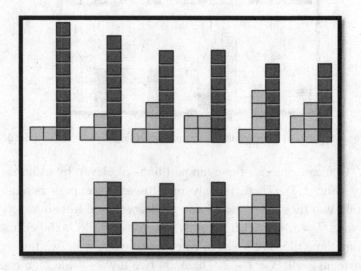

Fig. 5.7. Reduction to a problem with two towers when focusing on the tallest tower.

Task 5. *Find all ways to build three different size towers out of twelve square tiles and arrange them in the increasing order.*

Solution. Fig. 5.8 shows that three towers, arranged in the strictly increasing order, can be constructed in seven ways out of twelve tiles. Note that the towers are constructed through a system reflected in the following numeric interpretation of the diagram of Fig. 5.8:

$12 = 1 + 2 + 9, 12 = 1 + 3 + 8, 12 = 1 + 4 + 7, 12 = 1 + 5 + 6; 12 = 2 + 3 + 7,$

$12 = 2 + 4 + 6, 12 = 3 + 4 + 5.$

Indeed, because $12 - 1 = 11$ and $11 = 2 + 9 = 3 + 8 = 4 + 7 = 5 + 6$ (not including the case $11 = 1 + 10$ because of partitioning into unequal summands), there are four tri-towers with a single tile used for the smallest tower. Because $12 - 2 = 10$ and $10 = 3 + 7 = 4 + 6$ (not including $10 = 5 + 5$, $10 = 1 + 9$ and $10 = 2 + 8$ because of partitioning into unequal summands), there are two tri-towers with two tiles used for the smallest tower. Finally, because $12 - 3 = 9$ and $9 = 4 + 5$ is the only way to decompose 9 in two summands greater than 3 without regard to order, there is only one tri-tower with three tiles used for the smallest tower.

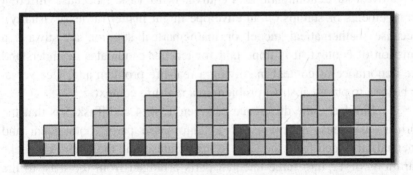

Fig. 5.8. Partition of twelve into three unequal parts focusing on the smallest part.

Another way to solve Task 5 is to put 12 tiles in three parts focusing on the largest part (thus interpreting the diagram of Fig. 5.8 differently). To this end, first note that it is not possible to have 11-tile and 10-tile high tri-towers because one needs at least three tiles to make two unequal towers. How many 9-tile high towers are there? Because $3 = 2 + 1$, there is only one tri-tower 9-tile high. Likewise, there is only one tri-tower 8-tile high because $4 = 3 + 1$ (not including $4 = 2 + 2$). There are two tri-towers 7-tile high because $5 = 4 + 1$ and $5 = 3 + 2$. There are two tri-towers 6-tile high because $6 = 5 + 1$ and $6 = 4 + 2$ (not including $6 = 3 + 3$). Finally, there is only one tri-tower 5-tile high because seven can be decomposed in two unequal integers smaller than five in only one way, $7 = 3 + 4$.

5.3 Activities with Towers Motivate Introduction of Algebraic Notation

The difference between Task 1 and Task 2 is that the former seeks all partitions of four with regard to the order of summands (e.g., considering $4 = 1 + 1 + 2$ and $4 = 2 + 1 + 1$ as different partitions) and the latter seeks all partitions of four without regard to the order of summands (i.e., not making distinction between the two partitions). Other examples of contexts different from the construction of towers can be presented. For instance, asking for a number of ways to spend \$4 by buying drinks priced \$1, \$2, \$3, and \$4 is equivalent to the query of Task 2. Asking for a number of ways to put a \$4-postage on an envelope by using stamps of the above four denominations is equivalent to Task 1 because different arrangements of stamps on an envelope might matter (at least visually). Because mathematical model or mathematical solution are always a function of context, it is important for teacher candidates to understand the importance of context in solving a real-life problem and, vice versa, to be able to put an abstract problem in a real-life context.

Similarly, the difference between Task 4 and Task 5 is that the former task seeks partitions of 11 into three parts (both equal and unequal) arranged from the least to the greatest, and the latter task seeks partitions of 12 into three unequal parts arranged from the least to the greatest. For example, the diagram of Fig. 5.6 might prompt the context of putting 11 people in three groups, equal or not. On the contrary, the diagram of Fig. 5.7 points at unequal groups only.

For the purpose of generalization, different contexts require using different notations. Let $P(n)$ and $p(n)$ be the number of partitions of n considered in the context of Task 1 ($n = 4$) and Task 2 ($n = 4$), respectively, that is, with and without the order of summands. Also, let $P(n, m)$ and $Q(n, m)$ be the number of partitions introduced in the context of Task 4 and Task 5, respectively, where the numbers 11 (Task 4) and 12 (Task 5) are replaced by n and the number 3 (Tasks 4 and 5) is replaced by m. In what follows, $P(n)$ and $p(n)$ will be referred to as partition of n with and without regard to the order of parts, respectively; $P(n, m)$ as partition of n into m parts and $Q(n, m)$ as partition of n into m non-equal parts, both without regard to order.

5.4 Ferrers-Young Diagrams

A representation of a positive integer as a sum of like numbers through an array of dots in which each addend is represented by a row or a column of dots is called Ferrers[9] diagram [Skiena, 1990]. To reflect possible activities which, in the context of Tasks 1 – 4, are appropriate for young children and their future teachers alike, the diagrams will be represented below using square tiles. Such diagrams (in which tiles are used instead of dots) are also called Young[10] diagrams [Stanton and White, 1986]. So, Figs 5.3 – 5.8 may be called Ferrers-Young diagrams representing partition of four into all (ordered) parts (Fig. 5.3), partition of four into unordered parts (Fig. 5.4, not counting towers shaded dark), partition of eleven into three parts (Fig. 5.5 and Fig. 5.6), and partition of twelve into three unequal parts (Fig. 5.7 and Fig. 5.8).

5.5 Recursive Definition of $P(n, m)$ Informed by Ferrers-Young Diagrams

One can use towers to develop mathematical definitions for the partitions $P(n, m)$ and $Q(n, m)$ to enable their numerical modeling for different n and m within a spreadsheet. In turn, a spreadsheet environment can then be used to develop grade-appropriate activities for "mathematically proficient students" [Common Core State Standards, 2010] and their future teachers that enable the discovery of interesting relationships among different kinds of partitions mentioned above and other mathematical concepts, including Pascal's triangle and the binomial coefficients, triangular numbers, difference equations and quadratic sequences as their solutions, and systems of linear equations.

---

[9] Norman Macleod Ferrers (1829–1908) – a British mathematician.

[10] Alfred Young (1873–1940) – a British mathematician.

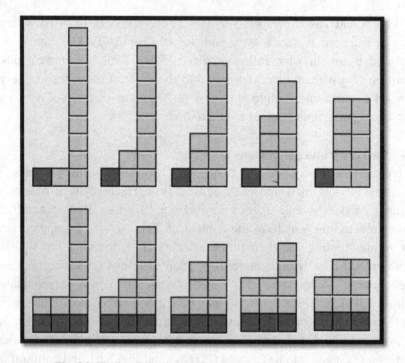

Fig. 5.9. Finding $P(11, 3)$ by reduction to two simpler problems.

There is yet another way to construct tri-towers (arranged from
the least to the greatest) out of eleven tiles by reducing the construction
to the following two cases. In the first case (the upper part of Fig. 5.9),
one has to construct tri-towers one element of which is a single tile. Such
towers can be constructed by using ten tiles to build two towers
(arranged from the smaller to the larger). That is, the problem of
constructing tri-towers is partially reduced to a simpler problem of
constructing bi-towers. In the second case (the bottom part of Fig. 5.9),
all the remaining tri-towers can also be constructed out of a smaller
number of tiles if one notes that eleven tiles may be diminished by three
tiles because, due to the absence of a one-story tower (they all have been
taken care in the first case), the first level in each of the tri-towers to be
constructed may be removed thereby leaving again tri-towers but to be
constructed out of a smaller (eight) number of tiles. In other words, the
first group of tri-towers is characterized by the number $P(10, 2)$, which is

equal to 5 as shown in Fig. 5.9, and the second group of tri-towers – by the number $P(8, 3)$, which is also equal to 5 as shown in Fig. 5.9. That is, $P(11, 3) = P(10, 2) + P(8, 3)$, where $10 = 11 - 1$, $2 = 3 - 1$, and $8 = 11 - 3$.

In general, the following recursive relation can be used to define $P(n, m)$

$$P(n, m) = P(n - 1, m - 1) + P(n - m, m). \tag{5.1}$$

In order for a recursive definition to be complete, one has to define the so-called boundary conditions (see also Chapter 7, section 7.5) that typically reflect common sense:

$$P(n, n) = P(n, 1) = 1 \text{ and } P(1, m) = 0 \text{ for } m > 1. \tag{5.2}$$

Indeed, there is only one way to construct n towers as well as a single tower out of n square tiles, and it is impossible to make two or more parts out of a single tile.

A	B	C	D	E	F	G	H	I	J	K	L	M	X
1													
2	m\n	1	2	3	4	5	6	7	8	9	10	11	
3	1	1	1	1	1	1	1	1	1	1	1	1	
4	2		1	1	2	2	3	3	4	4	5	5	
5	3			1	1	2	3	4	5	7	8	10	
6	4				1	1	2	3	5	6	9	11	
7	5					1	1	2	3	5	7	10	
8	6						1	1	2	3	5	7	
9	7							1	1	2	3	5	
10	8								1	1	2	3	
11	9									1	1	2	
12	10										1	1	
13	11											1	
24													
25	p(n)	1	2	3	5	7	11	15	22	30	42	56	

Fig. 5.10. Spreadsheet modeling of relations (5.1) and (5.2).

One can use the spreadsheet shown in Fig. 5.10 to model numbers $P(n, m)$ satisfying relations (5.1) and (5.2). The spreadsheet can be programmed as follows. To reflect initial conditions (5.2) the range [C3:W3] is filled with the units and the range [C4:C23] is filled with

(hidden) zeroes (in Fig. 5.10 only the range [C3:M13] is shown). In cell
D4 the spreadsheets formula
=IF(D$2-$B4<0,0,IF($B4-D$2=0,1,C3+INDEX($C4:C4,1,D$2-$B4)))
is defined and replicated across rows and down columns to cell W23. In
cell C2 the formula = SUM(C3:C23) is defined and replicated across row
25. In particular, the spreadsheet confirms that $P(10, 2) = P(8, 3) = 5$.
The meaning of the numbers in row 25 will be explained in the next
section.

5.6 Making Mathematical Connections

Let us now return to the diagram of Fig. 5.4. One can see that there are
five ways to partition four into non-negative integers without regard to
order; namely,
$$4 = 1+1+1+1 = 1+1+2 = 1+3 = 2+2 = 4+0.$$
At the same time, the non-dark towers in Fig. 5.4 can be seen as the
result of building (out of four tiles) four, three, two, and one towers, the
elements of which are arranged in the non-decreasing order. In other
words, these towers can be constructed, respectively, in $P(4, 4)$, $P(4, 3)$,
$P(4, 2)$, and $P(4, 1)$ ways. That is, the number of partitions of four in
whole number summands without regard to order, $p(4)$, can be found as
follows
$$p(4) = P(4, 4) + P(4, 3) + P(4, 2) + P(4, 1).$$

Likewise, the diagram of Fig. 5.5 shows seven ways to partition
five into whole number summands without regard to order. At the same
time, the non-dark towers in Fig. 5.5 can be seen as a result of building
(out of five tiles) one, two, three, four, and five towers the elements of
which are arranged in the non-decreasing order. In other words, these
towers can be constructed in $P(5, 1)$, $P(5, 2)$, $P(5, 3)$, $P(5, 4)$, and $P(5, 5)$
ways. That is,
$$p(5) = P(5, 1) + P(5, 2) + P(5, 3) + P(5, 4) + P(5, 5).$$
One can conjecture that, in general,
$$p(n) = P(n, 1) + P(n, 2) + ... + P(n, n). \tag{5.3}$$
That is, all the towers built out of n tiles and arranged in a non-
decreasing order can be put in n groups as follows: the first group
comprises a single, n-tile tall tower, the second group – all bi-towers, the

third group – all tri-towers, and so on. Indeed, $P(5, 1) = P(5, 4) = P(5, 5) = 1$, $P(5, 2) = P(5, 3) = 2$, and, due to (5.3), $p(5) = 1 + 2 + 2 + 1 + 1 = 7$.

Note that the array of numbers $P(n, m)$ resembles Pascal's triangle – a triangular array filled with integers (called binomial coefficients and expressed through combinations without repetitions introduced in Chapter 2), so that any element of the triangle is the sum of the elements immediately to its left in the same row and in the row above it. It is interesting to explore whether Pascal's triangle can be connected to a partitioning problem. In Fig. 5.10 we have $p(4) = 1 + 2 + 1 + 1 = 5$; in Fig. 5.11 we have $C_3^0 + C_3^1 + C_3^2 + C_3^3 = 1 + 3 + 3 + 1 = 8$. How can one interpret the entries of Pascal's triangle in term of partitions? For example, how can one partition four into two, three, or four ordered summands?

To answer the last question, consider the following diagram

$$\{1 \quad 1 \quad 1\} + \{1\} \quad \rightarrow \quad 4 = 3 + 1$$
$$\{1 \quad 1\} + \{1 \quad 1\} \quad \rightarrow \quad 4 = 2 + 2$$
$$\{1\} + \{1 \quad 1 \quad 1\} \quad \rightarrow \quad 4 = 1 + 3$$

in the left-hand side of which the partitioning of four into two summands is carried out by placing the plus sign between two groups of units (see also the diagram of Figure 3.11 in Chapter 3, section 3.5.2). The question to be answered is: How many ways a single plus sign can be put in three spaces that separate four units? The answer is C_3^1, that is, the number of ways to choose one object out of three objects. We have one plus sign because of two summands. We have three spaces because four units require three spaces to be separated. However, when four is partitioned into three summands, we have three spaces where two plus signs have to be placed and this can be done in C_3^2 ways. In general, a number n can be partitioned into m summands (counting partitions into the same but differently ordered parts) in C_{n-1}^{m-1} ways. Therefore, with regard to the order of summands, C_3^0 is the number of ways to partition four in one summand, C_3^1 — to partition four in two summands, C_3^2 — to partition four in three summands, and C_3^3 — to partition four in four summands. Now, one can see that the situation introduced in Task 1 could be resolved by connecting the array of numbers $P(n, m)$ to Pascal's triangle.

It follows from the well-known identity (e.g., Cuoco [2005]) $\sum_{m=0}^{n-1} C_n^m = 2^{n-1}$ and the fact that the largest number of plus signs that can be put between n units is equal to $n - 1$, that n can be partitioned in all differently ordered positive integer summands in 2^{n-1} ways. In other words, the equality $P(n) = 2^{n-1}$ holds true. In particular, when $n = 5$ (*cf.* Fig. 5.5) we have $P(5) = 2^4 = 16$. An alternative way of arriving at the same result by using two-color counters and tree diagrams is discussed in Chapter 3, section 3.5.2 and in more detail in [Abramovich, 2010a].

m\n	1	2	3	4	5	6	7	8
1	1	1	1	1	1	1	1	1
2		1	2	3	4	5	6	7
3			1	3	6	10	15	21
4				1	4	10	20	35
5					1	5	15	35
6						1	6	21
7							1	7
8								1
2^n	1	2	4	8	16	32	64	128

Fig. 5.11. Pascal's triangle.

5.7 Recursive Definition of $Q(n, m)$ Informed by Ferrers-Young Diagrams

A recursive definition can also be developed for the number of partitions of n into m unequal parts without regard to the order of the parts, that is, for $Q(n, m)$. The method of reduction to two simpler problems differs from the case for $P(n, m)$ when parts may be equal. Let us consider $n = 12$ and $m = 3$ already shown (non-recursively) in Fig. 5.8. Now, two cases need to be considered. In the first case (the upper part of Fig. 5.12), one can build a foundation out of three tiles and then augment each foundation by a bi-tower built out of nine tiles the elements of which are arranged in the strictly increasing order. Numerically, we have four such partitions of nine, namely, $9 = 1 + 8 = 2 + 7 = 3 + 6 = 4 + 5$. In the second case (the bottom part of Fig. 5.12), all the remaining tri-towers can also be constructed out of a smaller number of tiles by building on a three-tile foundation tri-towers constructed out of nine tiles the elements of which are arranged in the strictly increasing order. Numerically, we have three such partitions of nine, namely, $9 = 1 + 2 + 6 = 1 + 3 + 5 = 2 + 3 + 4$. That is, $Q(12, 3) = Q(9, 2) + Q(9, 3)$ where $9 = 12 - 3$ and $2 = 3 - 1$.

In general, similarly to the case of $P(n, m)$ which was also mediated by a Ferrers-Young diagram, all partitions of n into m distinct parts, $Q(n, m)$, can be put in two groups. The first group includes all such partitions with the number 1 present; the second group includes the remaining partitions. The number of partitions in the first group is equal to $Q(n - m, m - 1)$ because, in the language of tiles and towers, after partitioning $n - m$ tiles into $m - 1$ distinct parts it is possible to augment each $(m - 1)$-tower partition by an m-tile long base – this creates all partitions of n tiles into m distinct parts each of which has the first element equal to the number 1. Likewise, the number of partitions in the second group is equal to $Q(n - m, m)$ because after partitioning $n - m$ tiles into m distinct groups, it is possible to augment each m-tower partition with an m-tile long base – this creates all partitions of n tiles into m distinct parts none of which has a single element present.

This leads to the following recursive definition

$$Q(n, m) = Q(n - m, m - 1) + Q(n - m, m) \qquad (5.4)$$

where

$$Q(n, 1) = 1 \text{ and } Q(1, m) = 0 \text{ for } m > 1. \qquad (5.5)$$

Indeed, there is only one way to construct a single tower out of n square tiles and it is impossible to break one tile in two or more parts.

One can use the spreadsheet shown in Fig. 5.13 to model numbers $Q(n, m)$ satisfying relations (5.4) and (5.5). The spreadsheet can be programmed as follows. To reflect initial conditions (5.5), row 3 is filled with the units; column B from cell B4 is filled with (hidden) zeros; in cell C4 the spreadsheet formula

=IF(OR($A4-C$2>=0,$A3-B$2>=0),0,IF(AND($A4-C$2=0,$A3-B$2=0),1,INDEX($B3:B3,1,B$2-$A3)+INDEX($B4:B4,1,C$2-$A4))) is defined and replicated across rows and down columns to cell AN8. In particular; the modeling confirms that $Q(10, 3) = Q(7, 2) + Q(7, 3)$.

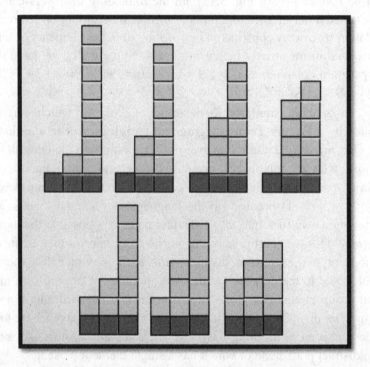

Fig. 5.12. Representing Eq. (5.4) in the case $n = 12$ and $m = 3$.

m\n	1	2	3	4	5	6	7	8	9	10	11	12	13	14	15	16	17	18	19	20	21	22	23	24	25	26	27	28	29	30	31
1	1	1	1	1	1	1	1	1	1	1	1	1	1	1	1	1	1	1	1	1	1	1	2	3	4	5	6	7	8	9	10
2			1	1	2	2	3	3	4	4	5	5	6	6	7	7	8	8	9	9	10	10	11	11	13	14	17	19	23	26	31
3						1	1	2	3	4	5	7	8	10	12	14	16	19	21	24	27	30	33	37	40	44	48	53	58	65	72
4										1	1	2	3	5	6	9	11	15	18	23	27	34	39	47	54	64	72	84	94	108	120
5															1	1	2	3	5	7	10	13	18	23	30	37	47	57	70	84	101
6																					1	1	2	3	5	7	11	14	20	26	35
7																												1	1	2	3
8																															

Fig. 5.13. Spreadsheet modeling of relations (5.4) and (5.5).

5.8 Connection to Triangular Numbers Opens a Window to a New Concept

What is the meaning of $Q(10, 4)$? The symbol stands for the number of partitions of 10 into 4 unequal summands. There is only one such partition, $10 = 1 + 2 + 3 + 4$; that is, $Q(10, 4) = 1$ (confirmed by the spreadsheet of Fig. 5.13). Furthermore, the number 10 is the smallest integer for which partition into four unequal summands is possible. It is the smallest because the sum starts with the smallest positive integer and it includes consecutive integers only. The sums of consecutive natural numbers starting from one are called triangular numbers (see Chapter 6, section 6.3.2 for more detail) and were also mentioned in the footnote to section 5.2 of this chapter. Likewise, because $1 + 2 + 3 = 6$, the number 6, being a triangular number, is the smallest integer for which a partition into three unequal integers is possible. For example, the equality $10 = 1 + 3 + 6$ mentioned in section 5.2 represents a partition of 10 into three unequal integers. Such partition is only possible for integers $n \geq 6 = \dfrac{3(3+1)}{2}$. Indeed, it is not possible to represent the number 5 as a sum of three unequal natural numbers, that is, $Q(5, 3) = 0$.

Generalizing from these observations yields the relation

$$Q(n, m) = 0 \text{ for all } n < \frac{m(m+1)}{2}.$$

For example, $Q(44, 9) = 0$ because $44 < \dfrac{9 \cdot 10}{2}$, yet $Q(45, 9) = 1$ because 45 is the triangular number of rank 9; that is, $45 = 1 + 2 + 3 + \ldots + 9$.

The data presented in the spreadsheet of Fig. 5.13 shows that for some values of n the largest value of m in $Q(n, m)$ is equal to three. For example,

$$Q(9, 3) = Q(13, 4) = Q(18, 5) = Q(24, 6) = Q(31, 7) = 3. \qquad (5.6)$$

That is, the number of partitions of 9 in 3 unequal parts is the same as the number of partitions of 13 in 4 unequal parts and it is the same as the number of partitions of 18 into 5 unequal parts, and so on. Can this observation, that is, relations (5.6), be generalized? To answer this question, consider the sequence

$$9, 13, 18, 24, 31, \ldots \qquad (5.7)$$

along with the sequence of triangular numbers that starts with the third
triangular number

$$6, 10, 15, 21, 28, \ldots . \tag{5.8}$$

One may note that by increasing each term of sequence (5.8) by three,
the corresponding term of sequence (5.7) results. As sequence (5.8) has

the form $x_n = \dfrac{(n+2)(n+3)}{2}$ for $n = 1, 2, 3, \ldots$, sequence (5.7) can be

written in the form $y_n = \dfrac{(n+2)(n+3)}{2} + 3$. Consequently, relations (5.6)

can be generalized as follows

$$Q(\frac{(n+2)(n+3)}{2} + 3, n+2) = 3. \tag{5.9}$$

Indeed, (5.6) results from (5.9) for $n = 1, 2, \ldots, 5$. Noting that $p(3) = 3$,
relation (5.9) can be interpreted as follows: the number of partitions of
the number 3 into integer parts without regard to order (these partitions
are $3 = 1 + 1 + 1, 3 = 1 + 2, 3 = 3$) is equal to the number of partitions of,
increased by three, the triangular number of the rank $n + 2$ into $n + 2$
unequal parts. For example, 10 is the triangular number of rank **four** and
the number 13 (= 10 + 3) can be partitioned into **four** unequal positive
integers in three ways: $13 = 1 + 2 + 3 + 7$, $13 = 1 + 2 + 4 + 6$, and $13 = 1
+ 3 + 4 + 5$.

Likewise, using the extension of the spreadsheet of Fig. 5.13 the
following relations can be developed through empirical induction

$$Q(\frac{(n+3)(n+4)}{2} + 4, n+3) = 5, \; Q(\frac{(n+4)(n+5)}{2} + 5, n+4) = 7,$$

$$Q(\frac{(n+5)(n+6)}{2} + 6, n+5) = 11, \; Q(\frac{(n+6)(n+7)}{2} + 7, n+6) = 15.$$

But just as in (5.9), the right-hand sides of the last four equalities are,
respectively, $p(4)$, $p(5)$, $p(6)$, and $p(7)$ as shown in Fig. 5.10. This
observation leads to another generalization:

$$Q(\frac{(n+k-1)(n+k)}{2} + k, n+k-1) = p(k), \quad n, k = 1, 2, 3, \ldots . \tag{5.10}$$

Finally, relation (5.10) can be conceptualized in the following form:

> *The number of partitions of k into integer parts without regard to*
> *order is equal to the number of partitions of the, increased by k,*

triangular number of the rank $n+k-1$ into $n+k-1$ unequal parts arranged from the least to the greatest for any positive integer value of n.

In particular, substituting $k = 4$ and $n = 1$ into (5.10) yields

$$Q(\frac{4 \cdot 5}{2}+4, 4) = Q(14, 4) = p(4) = 5 \text{ and}$$

$$14 = 1+2+3+8 = 1+2+4+7 = 1+2+5+6 = 1+3+4+6 = 2+3+4+5.$$

As shown in Fig. 5.14, the number of partitions of the number 4 into three parts without regard to order (the towers at the top) is equal to the number of unordered partitions of the number 14 into four unequal parts (the towers at the bottom).

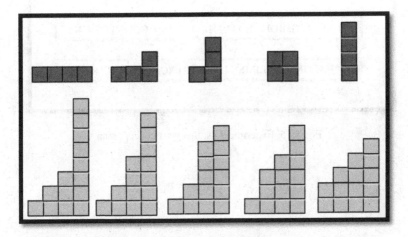

Fig. 5.14. Confirming theory through Ferrers-Young diagrams.

To conclude, note that this chapter, using the theory of partitions as background, demonstrated how the use of concrete materials (the first-order symbols) can become a genesis for the creation of theory via the use of the W^4S principle and spreadsheet modeling. Through all the steps of abstraction and generalization, theoretical developments were verified using special cases, both through the use of the first-order symbols and the second-order symbolism. Finally, it was demonstrated how the power of technology allowed for uncovering rather intricate mathematical connections enabling different concepts to come together in forming a conceptually robust mathematical proposition. A concept map of this approach is shown on Fig. 5.15.

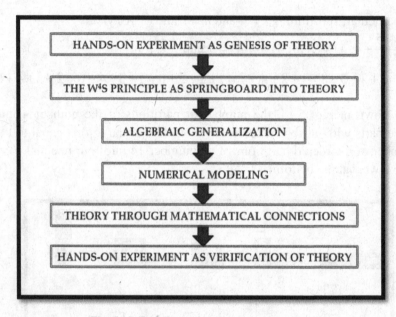

Fig. 5.15. Experiment vs. theory: a concept map.

Chapter 6

Hidden Curriculum of Mathematics Teacher Education

6.1 Introduction

The pedagogy of this chapter incorporates the hidden mathematics curriculum framework [Abramovich and Brouwer, 2006; Abramovich, 2014]. This learning framework is based on the notion that many seemingly disconnected mathematical activities and problems scattered across the K-12 mathematics curriculum are, in fact, connected through a common conceptual structure which is *hidden* from learners because of its intrinsic complexity. By the same token, many seemingly routine tasks, when explored beyond the boundaries of the traditional curriculum, can be used as windows to big ideas obscured in the curriculum. In the context of teacher education, such extended explorations require a certain level of mathematical competence and intellectual courage on the part of the instructor. Furthermore, the complex nature of explorations requires the development of learning environments conducive to revealing hidden mathematical concepts to teacher candidates in a pedagogically appropriate format.

Utilizing the hidden mathematics curriculum framework in a technological paradigm, when technology is understood broadly to include both computational and hands-on tools, provides significant opportunities to enhance the conceptual component of mathematics teacher education. For example, through spreadsheet modeling, traditionally difficult mathematical ideas can become embedded in this powerful digital tool enabling easy access to these ideas through appropriate pedagogical mediation [Abramovich and Sugden, 2008]. Likewise, modeling with manipulative materials allows one to create isomorphic relationships among the first-order symbols and the second-order symbolism. Such mediation occurs in the social context of competent guidance provided by the instructor who serves to teachers as 'a more knowledgeable other' [Vygotsky, 1978]. The transactional nature of knowledge acquisition, being the primary thesis of Vygotskian educational psychology, is also reflected in Freudenthal's [1983] interpretation of the didactical phenomenology of mathematics – "a way to show the teacher the places where the learner might step into the learning process of mankind" (p. ix). This combination of Vygotskian tradition to see teaching as assisted performance [Tharp and Gallimore, 1988] and Freudenthal's [1983] perspective on mathematics learning as the advancement of culture of mankind provides a theoretical foundation for the hidden mathematics curriculum framework. In this chapter, proceeding from a mundane task about adding consecutive natural numbers, it will be demonstrated how this framework can facilitate informed entries into a mathematical culture for elementary teacher candidates.

6.2　The Basic Task

Consider the following task adopted from [Van de Walle, 2001, p. 66]:

> *Find all the ways to add consecutive natural numbers*
> *to reach sums not greater than 20.*

This task, included in a textbook for future teachers of elementary and middle school mathematics, provides a conceptually rich activity having a flavor of an open-ended exploration using the operation of addition as background. (For some reason, the task was excluded from recent editions of the book). One of the main didactic values of open-ended mathematical tasks is their either explicit or implicit connection to big ideas, a rich conceptual structure, and the possibility of using such tasks with students of different mathematical abilities, interests, and grade levels. As will be shown below, this 'addition' task, indeed, enables multiple extensions and degrees of complexity in problem solving that can be used with young and not so young mathematics learners using a variety of tools: manual, computational and symbolic.

　　Typically, teacher candidates have no difficulty with the task and eagerly volunteer to demonstrate their well-mastered procedural skills. These skills are essentially automatic, whereas the task can be used to assess and foster one's creative thinking in mathematics. Recognizing this distinction is important because automatism and creativity have been observed as contradictory features of mathematical pedagogy: "sources of insight can be clogged by automatisms" [Freudenthal, 1983, p. 469]. Therefore, without proper advice, teacher candidates may get an erroneous impression that the main goal of this task is to support practicing addition of one-digit numbers (except the case $9 + 10 = 19$) rather than using an easy mathematical content for developing what is often referred to as deep understanding of the subject matter. Such understanding stems from the pedagogy of reflective inquiry described by Dewey [1933] as a problem-solving method that blurs the distinction between knowing and doing. In the mathematics classroom, reflective inquiry reduces the dichotomy between knowing mathematical content and understanding how to teach it. That is, by intertwining mathematical content and methods of teaching mathematics one can demonstrate how the diversity of the methods unfolds through the appropriate reflection on the content. It is the appreciation of this diversity that has the potential to change elementary teacher candidates' procedural perception of mathematics and routine methods of teaching the subject matter to young children.

Another goal of intertwining the content and the methods is to develop the awareness of one of the most characteristics features of mathematical classroom discourse: a student can naturally ask simple yet challenging questions about seemingly mundane and apparently well-understood concepts. Such a goal has major implications for the advancement of school mathematics instruction: in open-ended problematic situations, encouraged by current standards for teaching mathematics worldwide [Common Core State Standards, 2010; National Council of Teachers of Mathematics, 2000; Conference Board of the Mathematical Sciences, 2012; Advisory Committee on Mathematics Education, 2011; Department for Education, 2013; Ministry of Education Singapore, 2012; National Curriculum Board, 2008; Ontario Ministry of Education, 2005] students are likely to ask questions to which even knowledgeable teachers often do not have immediate answers. This viewpoint remains true for all levels of K-12 mathematics education. An extreme, yet classic example of that kind is the following question: Is it possible to extend the partitioning of a square into the sum of two squares (Pythagorean triples) to include similar representations for higher powers? For example, in terms of the first-order symbols, is it possible to decompose a cube built out of unit cubes into two smaller cubes? (As shown in Fig. 6.11 below such decomposition is possible for squares). It took some 350 years of efforts by many brilliant minds before the negative answer to this question, known as Fermat's Last Theorem, was finally found [Wiles, 1995]. There are still quite a few simply posed questions yet to be answered, among them: Can any even number greater than two be decomposed into the sum of two prime numbers (that is, numbers with exactly two different divisors)? For example, $4 = 2 + 2$, $6 = 3 + 3$, $8 = 3 + 5$, and so on. This easy-to-understand statement is known as Goldbach's Conjecture[11] and it was verified computationally for very large numbers. Therefore, it is important, especially for elementary teachers, to change their beliefs in the simplicity of mathematics they are to teach and move away from the absolute trust in the pedagogy of pure memorization of rules, formulas, and procedures towards the pedagogy that supports the dual intensity shift in classroom practices to enable both practicing and comprehending mathematics by students [Common Core State Standards, 2010].

6.3 Background Information: Triangular and Trapezoidal Numbers

6.3.1 *Activities*

This section begins with the description of several classroom activities aimed at the introduction of the concept of the sum of consecutive natural numbers. To this end, consider a situation when 19 students were divided into four groups with the number of students in each group being as close to each other as possible. This resulted in groups 1, 2 and 3, five students in each, and group 4 of four students[12]. Each group was assigned a different task as follows.

---

[11] Christian Goldbach (1690–1764) – a German mathematician.

[12] Put another way, 19 was divided by 4 yielding 4 as the quotient and 3 as the remainder: $19 = 4 \cdot 4 + 3 = 3 \cdot (4+1) + 4 = 3 \cdot 5 + 4$. That is, whereas 19 students can be arranged in four groups without regard to order in many ways (54, to be exact – see the value of $P(19, 4)$ in the spreadsheet of Fig. 7.10, Chapter 7) but only one arrangement represents the division of 19 by 4, that is, $19 = 5 + 5 + 5 + 4$.

Task for group 1. Find the number of ways two students can be selected from your group to report to the class on the group work.

Task for group 2. Find the number of handshakes that can be made among the group members when shaking each other's hand only once.

Task for group 3. Find all ways to arrange the group members into four subgroups.

Task for group 4. The students in this group are Al, Barb, Chris, and Drew. They were asked to do the following. Al claps one time and invites Barb to join him – they both clap simultaneously one time each and then invite Chris to join them. The three clap simultaneously one time each and then invite Drew to join them. Finally, the four clap in the same way. After that, they are asked to count the total number of claps that took place.

6.3.2 *Solutions to the tasks*

Solution reported by group 1. The students' initials are A, B, C, D, and E. They reported that four pairs can be formed with A, namely: (A, B), (A, C), (A, D), and (A, E); three pairs can be formed with B but without A, namely: (B, C), (B, D), and (B, E); two pairs can be formed with C but without either A or B, namely: (C, D) and (C, E); finally, the last pair is (D, E). Counting the pairs through the rule of sum (Chapter 2, section 2.2) yields the sum $1 + 2 + 3 + 4$. In other words, A can form four pairs, B can form three new pairs, C can form two new pairs, and D and E can form one new pair.

The above reasoning can be generalized to $n + 1$ students who can, thereby, be paired in $1 + 2 + 3 + \ldots + n$ ways. Indeed, if n students can be paired in $S(n)$ ways, then, with the addition of another student, n new pairs can be formed (pairing this new student with each of the other n students). That is, the number of pairs can be defined recursively through the formula

$$S(n+1) = S(n) + n, \, S(1) = 0, \, n = 1, 2, 3, \ldots .$$

We have n relations

$$S(2) = 1, S(3) = S(2) + 2, S(4) = S(3) + 3, ..., S(n+1) = S(n) + n,$$

adding which and then cancelling out equal terms in the left- and right-hand sides of the combined equation

$$S(2) + S(3) + S(4) + ... + S(n+1) = 1 + S(2) + 2 + S(3) + 3 + ... + S(n) + n$$

yield

$$S(n+1) = 1 + 2 + 3 + ... + n.$$

Alternatively, the last formula can be proved by the method of mathematical induction (though typically not familiar to elementary teacher candidates): the base clause $S(2) = 1$ is obvious; assuming $S(n) = 1 + 2 + 3 + ... + n - 1$ (inductive clause) yields the inductive transfer from n to $n + 1$, namely, $S(n + 1) = S(n) + n = 1 + 2 + 3 + ... + n$.

Solution reported by group 2. In counting handshakes, this group (having same initials as the members of group 1) used reasoning similar to the one demonstrated through the selection of two students out of five. Indeed, A did four handshakes and egressed, B did three handshakes and egressed, C did two handshakes and egressed; finally, D and E shook hands. Therefore, using, once again, the rule of sum, the expression $1 + 2 + 3 + 4$ emerged, now representing the number of handshakes among the five students. As above, one can prove that the number of non-repeated handshakes among $n + 1$ students is equal to the sum $1 + 2 + 3 + ... + n$.

Note that a new method of counting handshakes (or pairs, for that matter) can lead to a closed formula for the sum of the first n natural numbers. Such counting in the case of five students is shown in Fig. 6.1 using the tree diagram technique introduced in Chapter 2 (section 2.3). Because a tree diagram shows different orders in which objects are selected, in the trees of Fig. 6.1 the handshakes are counted twice. For example, on the first two trees (and only there) we see the handshakes (A, B) and (B, A). Likewise, on the last two trees (and only there) we see the handshakes (D, E) and (E, D). The five trees represent five groups of four pairs of students; that is, the product $5 \cdot 4$ is twice the number of handshakes among the five students. Recall that a tree diagram develops through the rule of product: the first member of a pair can be selected in five ways and, following this selection, the second member of this pair can be selected in four ways; therefore, an ordered pair of handshaking students can be selected in $5 \cdot 4 = 20$ ways. But although the order in which students extend hands to each other is immaterial, the tree diagrams of Fig. 6.1 show different orders: e.g., the branches AB and BA on the first two trees represent identical handshakes. Therefore, the total number of non-ordered pairs of handshaking students is half the above product, that is, 10. Generalizing this new way of counting handshakes to $n + 1$ students yields the product $n(n+1)$ as twice the number of handshakes among $n + 1$ people; that is, $2 \cdot (1+2+3+...+n) = (n+1)n$ whence

$$1 + 2 + 3 + ... + n = \frac{n(n+1)}{2} . \tag{6.1}$$

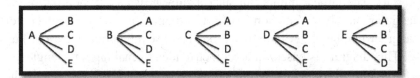

Fig. 6.1. The tree diagram counts handshakes twice.

Solution reported by group 3. Once again, the students' initials are A, B, C, D, and E. It was reported that there are *four* ways to pair A with other four students, with the remaining three students forming three groups; that is,

(A, B), C, D, E; (A, C), B, D, E; (A, D), B, C, E; and (A, E), B, C, D.

Next, there are *three* ways for A to form a one-person group while pairing B with the remaining three students: A, (B, C), D, E; A, (B, D), C, E; and A, (B, E), C, D. There are *two* ways for A and B to form two one-person groups while pairing C with the remaining two students: A, B, (C, D), E; and A, B, (C, E), D. Finally, there is only *one* way for A, B, and C to form three one-person groups while pairing D with E: A, B, C, (D, E). Therefore, once again, we have the sum $1 + 2 + 3 + 4$ as the total number of ways that four groups can be created out of five people. In general, the total number of n groups made out of $n + 1$ people is equal to the sum of the first n natural numbers, $1 + 2 + 3 + ... + n$.

Solution reported by group 4. Once again, we have four claps for Al, three claps for Barb, two claps for Chris, and one clap for Drew; that is, the total of $1 + 2 + 3 + 4$ claps. In general, there are $1 + 2 + 3 + ... + n$ claps among n people clapping as described above.

As we can see, different (socially oriented) counting situations, being structured by the rule of sum, lead to the need to find the sum of the first n natural numbers, $1 + 2 + 3 + ... + n$. These sums are called triangular numbers (already mentioned in Chapter 5, section 5.8). The term comes from antiquity when numbers were associated with geometric shapes in order to facilitate one's comprehension of their abstract meaning [Smith, 1953]. In Vygotskian terms, in the antiquity, the second-order symbolism was strongly associated with the first-order symbols. The largest number in such a sum, n, is called the rank of the triangular number. Relation (6.1) represents a closed formula for triangular numbers[13]. Alternatively, one can use a diagram shown in Fig. 6.2. This diagram is based on a geometric idea of making a rectangle out of two identical triangles. By modeling the sum $1 + 2 + 3 + ... + n$ in the form of a staircase made out of blocks, and then augmenting it with the identical (but flipped over) staircase, one has a rectangular structure which is n-block high and $(n + 1)$-block wide. Counting the number of blocks within the rectangle yields $n(n + 1)$ blocks. Taking half of this quantity yields formula (6.1) which now can be used to generate the sequence of triangular numbers $1, 3, 6, 10, 15, 21, ...$.

Triangular numbers $t_n = 1 + 2 + 3 + ... + n$ can also be defined using the recursive formula

$$t_n = t_{n-1} + n \tag{6.2}$$

where t_{n-1} is the triangular number of rank $n - 1$. Formula (6.2) can be given several contextual interpretations. For example, the total number of claps among n people is equal to the total number of claps among $n - 1$ people plus the number of claps when n people do one clap each. Likewise, $n + 1$ people can select one person to do n handshakes with everyone and after that, the remaining n people will do t_{n-1} handshakes.

---

[13] See Luchins and Luchins [1970b] for an interesting account of presenting this summation method "to fifth and six grade children in a slum neighborhood that used so-called drill methods of teaching arithmetic" (p. 95) by Max Wertheimer (1880-1943) – one of the founders of Gestalt psychology [Ellis, 1938].

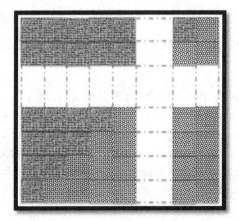

Fig. 6.2. Making rectangle out of two triangles.

6.3.3 *Trapezoidal numbers*

In order to motivate the summation of consecutive natural numbers starting from a number different from one, the task for group 4 can be modified as follows.

A modified task for group 4. A student named Eva was late for the class and joined group 4 after the task was explained. She was confused with the task and immediately clapped with Al not allowing the latter to do the first single clap, then Barb clapped jointly with Al and Eva and so on, until all five students clapped together. How many claps did take place?

Solution reported by the modified group 4. This time, rather than counting individual claps, the students counted claps at each step. This yielded the sum $2 + 3 + 4 + 5$. Other obvious modifications are possible to have sums that start with 3, 4, and so on (depending on how many students were late for class). In general, if $k - 1$ students were late for class and clapped simultaneously with the first student, the sum $k + (k + 1) + (k + 2) + ... + n$ can be used to count the number of claps. Here $n = (k-1) + m$, where m is the number of students not being late for class; that is, $m = n - k + 1$. Due to formula (6.1), one can write

$$k + (k+1) + ... + n = (1 + 2 + ... + k - 1 + k + ... + n) - (1 + 2 + ... + k - 1)$$
$$= \frac{n(n+1)}{2} - \frac{(k-1)k}{2} = \frac{n^2 - k^2 + n + k}{2} = \frac{(n+k)(n-k+1)}{2}.$$

The equality $N = k + (k + 1) + (k + 2) + ... + n$ is called a trapezoidal representation of a positive integer N with $n - k + 1$ rows, the latter number being the number of addends in the sum that represents N; in other words, a number that can be represented through a sum of consecutive natural numbers is called a trapezoidal number. When $k = 1$, a trapezoidal number turns into a triangular number. That is, a trapezoidal number may be considered a generalization of a triangular number. As will be shown below, in some cases the sum $k + (k + 1) + ... + n$ can be rearranged into a like sum with other first and last terms, thereby indicating that some integers may have more than one trapezoidal representation. For example, $9 = 4 + 5 = 2 + 3 + 4$. At the same time, the triangular representation of an integer, if it exists, is unique. This is quite obvious because the first term in the triangular representation of an integer is always one. At the same time, some numbers are not only non-triangular numbers, but also non-trapezoidal numbers. For example, $4 > 1 + 2$ and $4 < 2 + 3$; that is, the number 4 cannot be reached by adding consecutive natural numbers.

It follows from the formula

$$k + (k+1) + ... + n = \frac{(n+k)(n-k+1)}{2} \qquad (6.3)$$

that the numbers $n + k$ and $n - k + 1$ have to be of different parity to allow for the right-hand side of (6.3) to be an integer as its left-hand side may not be anything but integer. Regardless of formula (6.3), one can show that the difference $(n+k) - (n-k+1) = 2k - 1$ is an odd number (a double diminished by one), something that can only be true when the minuend and the subtrahend are of different parity. That is, if $n + k$ is an even number, then $n - k + 1$ is an odd number and vice versa. This relationship between $n + k$ and $n - k + 1$ will be used below (section 6.9) to suggest that if twice an integer does not have factors of different parity, it cannot be represented through a sum of consecutive natural numbers.

As a useful practice in formal reasoning using the language of the first-order symbols, one can prove that both the sum and the difference of two integers are of the same parity and, thereby, the integers $n + k$ and $n - k + 1$ are of different parity. To this end three cases need to be considered: (i) n and k are both even; (ii) n and k are both odd; (iii) n and k are of different parity. In case (i), adding or subtracting two sets of objects with no singles in each set yields a set with no singles (just imagine that we either add pairs to pairs or take away pairs from pairs). In case (ii), adding two sets of objects with a single in each set enables the singles to form a pair; when subtracting (that is, taking away) – the singles disappear. Finally, in case (iii), adding to a set of pairs another set with a single yields a set with a single; taking away a single from a set of pairs requires breaking one pair thus creating a new single. Therefore, because $n + k$ and $n - k$ are of the same parity, $n + k$ and $n - k + 1$ are of different parity.

6.4 Conceptually Oriented Discussion of the Basic Problem

The first conceptually oriented discussion of the basic problem of finding all ways to represent an integer through a sum of consecutive natural numbers may deal with different systematic ways of finding the representations sought. One way is to find all representations of two consecutive numbers, then of three, then of four, and so on. The second way is to group them according to the first addend (the top row in a trapezoidal representation). Yet, the third way is to use a brute force approach of partitioning integers into the sums of consecutive natural numbers. The main purpose of having multiple ways of finding all the representations is to show different ways of systematic reasoning each of which may be called a proof. In other words, reasoning through a system provides a formal demonstration that all the sums sought were found.

6.4.1 *Grouping the sums by the number of addends*

One strategy (used by teacher candidates) of introducing systematic reasoning in the creation of the sums is to put them in groups by the number of addends. To this end, note that there are nine sums with two addends:

$$1 + 2 = 3, 2 + 3 = 5, 3 + 4 = 7, 4 + 5 = 9, 5 + 6 = 11, 6 + 7 = 13,$$

$$7 + 8 = 15, 8 + 9 = 17, 9 + 10 = 19;$$

there are five sums with three addends:

$$1 + 2 + 3 = 6, 2 + 3 + 4 = 9, 3 + 4 + 5 = 12, 4 + 5 + 6 = 15,$$
$$5 + 6 + 7 = 18;$$

there are three sums with four addends:

$$1 + 2 + 3 + 4 = 10, 2 + 3 + 4 + 5 = 14, 3 + 4 + 5 + 6 = 18;$$

and there are two sums with five addends:

$$1 + 2 + 3 + 4 + 5 = 15, 2 + 3 + 4 + 5 + 6 = 20.$$

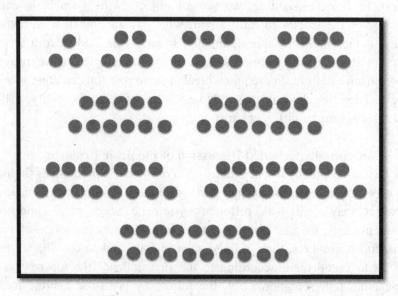

Fig. 6.3. Nine trapezoids with two rows.

Because the smallest sum with six addends is 21 – the sixth triangular number – in all, there are 19 different ways of representing integers not greater than 20 through the sums of consecutive natural numbers. In terms of the first-order symbols (Figs 6.3–6.6), there are 19 different trapezoids (including triangles interpreted as trapezoids with the top row equal one) that can be constructed out of 20 counters. One can immediately recognize that 19 is one smaller than 20 and ask whether the total number of sums can be determined without actually creating the sums. This is one of the questions when a teacher may say, 'A good question, let us think together'. When thinking together to connect 20 and 19 through the relation $19 = 20 - 1$, the latter can be immediately defied generality by replacing 20 by, say, 10 (for which seven sums were found above), thereby providing a counter-example ($7 \neq 10 - 1$) to a naïve conjecture. Yet, conjecturing must be encouraged in the classroom in order to learn different ways to confirm of defy conjectures. Some kind of an answer to the question about finding the number of representations without actually creating the sums will be developed throughout this chapter.

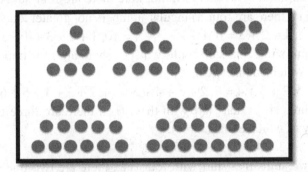

Fig. 6.4. Five trapezoids with three rows.

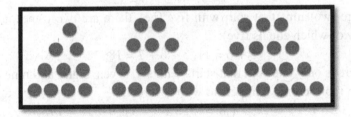

Fig. 6.5. Three trapezoids with four rows.

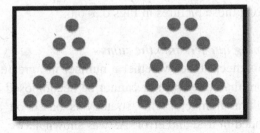

Fig. 6.6. Two trapezoids with five rows.

6.4.2 *Grouping the sums by the first addend*

Another strategy that makes it possible to introduce a system into finding the sums of consecutive natural numbers is to put them in groups organized by the first addend. Teacher candidates use this strategy also during a classroom presentation, although some are curious as to why we need multiple strategies. Considering the development of multiple strategies as a method of teaching mathematics, we will see how the diversity of methods can motivate the emergence of a new mathematical content to explore. With this in mind, note there are four sums that start with one (i.e., there are four triangular numbers not greater than 20):

$$1+2=3, 1+2+3=6, 1+2+3+4=10, 1+2+3+4+5=15;$$

there are four sums that start with two (i.e., there are four trapezoids the top row of which equals two):

$$2+3=5, 2+3+4=9, 2+3+4+5=14, 2+3+4+5+6=20;$$

there are three sums that start with three (i.e., there are three trapezoids the top row of which equals three):

$$3+4=7, 3+4+5=12, 3+4+5+6=18;$$

there are two sums that start with four (i.e., there are two trapezoids the top row of which equals four):

$$4+5=9, 4+5+6=15;$$

there are two sums that start with five (i.e., there are two trapezoids the top row of which equals five):

$$5+6=11, 5+6+7=18;$$

and there is only one sum that starts with six, seven, eight, and nine (each of them representing a trapezoid with two rows):

$$6+7=13; 7+8=15; 8+9=17; 9+10=19.$$

Once again, in all, 19 sums have been found using a different reasoning strategy. The corresponding trapezoidal representations can be recognized among those pictures in Figs 6.3–6.6.

6.4.3 *Partitioning integers into the sums*

Finally, one can check to see whether a number not greater than 20 can (perhaps in more than one way) or cannot be decomposed into a sum of consecutive natural numbers. This strategy is based on a brute force approach grounded in trial and error. As was shown above, in the case of impossibility for the number 4, one has to show that by trying all possible ways of making the sums one can, indeed, miss certain numbers.

$3 = 1 + 2, 5 = 2 + 3, 6 = 1 + 2 + 3, 7 = 3 + 4, 9 = 4 + 5, 9 = 2 + 3 + 4,$
$10 = 1 + 2 + 3 + 4, 11 = 5 + 6, 12 = 3 + 4 + 5, 13 = 6 + 7,$
$14 = 2 + 3 + 4 + 5, 15 = 7 + 8, 15 = 4 + 5 + 6, 15 = 1 + 2 + 3 + 4 + 5,$
$17 = 8 + 9, 18 = 3 + 4 + 5 + 6, 18 = 5 + 6 + 7, 19 = 9 + 10,$
$20 = 2 + 3 + 4 + 5 + 6.$

It should be noted that finding two partitions for the number 9 or three partitions for the number 15 is more difficult than through using either of the first two strategies. Indeed, one can ask why there are only three partitions for 15. The answer 'because we could not find more' is not a good one. At the same time, the numbers 4, 8 and 16 could not be decomposed into the sums of consecutive natural numbers. Indeed, $2 + 3 + 4 > 8$ and $3 + 4 < 8$; one can be asked to explain why the two inequalities are sufficient to defy 8 as a trapezoidal number and find the smallest number of inequalities that defy 16 as a trapezoidal number. Also, the third strategy (through which, once again, 19 sums have been found) can be used to confirm the correctness of the results obtained through the first two strategies.

6.5 Discourse Motivated by Multiple Ways of Creating Sums of Consecutive Natural Numbers

6.5.1 *Clarifying the meaning of the word special in the context of arithmetic*

Another purpose of encouraging multiple ways of systematic reasoning in demonstrating that all possible representations of natural numbers not greater than 20 through the sums of consecutive natural numbers have been found is to show how one can go beyond the immediate goal of the task in order to highlight mathematical ideas that underlie each demonstration. For example, in the context of summation presented in section 6.4.1, one can be asked: What is special about the sums of two consecutive natural numbers? What is special about the sums of three consecutive natural numbers? What is special about the sums of four

consecutive natural numbers? Here, the meaning of the word *special* has
to be explained because it is not universally understood in connection
with numbers. What do we mean by calling some numbers special? As
known from the history of mathematics, a physical manipulation of
objects and visual argumentation regarding the relationship among the
objects led to the need for names describing *specific* properties of
numbers. It is due to a discourse that unfolds through reflective inquiry
that mathematical meaning of the word special becomes more familiar to
teacher candidates. This, in turn, creates greater self-confidence in their
abilities to teach mathematics through conversation [Haroutunian-
Gordon and Tartakoff, 1996].

The importance of developing such a confidence can be seen in
the following reflective comment by an elementary teacher candidate
who admitted, "*my biggest fear . . . is that one bright student might
figure something out in a remarkable way and I won't be able to
recognize it and keep the student interested enough to be excited about
the next lesson.*" Facilitating teacher candidates' learning hidden
mathematical ideas in computationally supported environments helps
alleviate these kinds of concerns, especially at the elementary level. By
experiencing the association of traditional problem-solving activities
with rather sophisticated mathematical contexts, teacher candidates
develop greater understanding of, and confidence in, presenting these
activities to their own students. As another pre-teacher noted, "*... it is
hoped that I can implement skills learned in this course to ... challenge
students to see the relationships.*" In such a way, developing future
teachers' self-confidence in teaching mathematics through revealing non-
trivial ideas of the conventional curriculum is the major learning
objective of a course that incorporates the notion of the hidden
mathematics curriculum.

6.5.2 *The first encounter with a special property of the sums of two addends*

Once it is recognized that the sums of two consecutive natural numbers
are consecutive odd numbers, the next series of questions can be asked:
Why is it so? How can this be explained in terms of the relationship
between the first-order symbols and the second-order symbolism? Why

are these odd numbers consecutive ones? It is the practice of articulating answers to these questions that is important for teacher candidates' conceptual development.

There are two parts in answering the above questions about odd sums: why are they odd and why are they consecutive odd numbers? Seeing the sum of two consecutive natural numbers as a double plus one explains the meaning of an odd number in terms of the first-order symbols (Fig. 6.7). For example, consider the equality $1 + 2 = 3$. What do we see here? In other words, how can this equality be represented through the first-order symbols? Fig. 6.7 (far left) shows such a representation: a pair of counters (objects) and a single counter (object). That is, the sum is, indeed, a double plus one. This characterization does not change as one moves from the sum $1 + 2$ to that of $2 + 3$. On a physical level, we bring a pair of objects, thus moving from one odd number to the next odd number. This situation does not change as 2 is replaced by n and 3 is replaced by $n + 1$, making the sum of any two consecutive natural numbers equal $2n + 1$ (i.e., double of n plus one). One can see how the task opens a window from arithmetic to algebra. Put another way, the average of any two consecutive natural numbers is never a natural number.

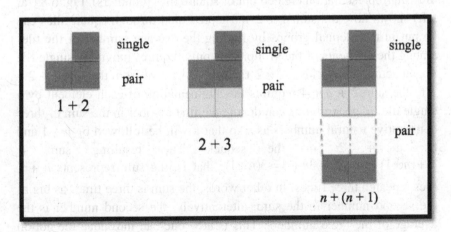

Fig. 6.7. The sum of two consecutive integers is always an odd number.

6.5.3 *Moving from novice to expert practice in revealing special properties*

The next activity may be to explore the properties of the sums of three consecutive natural numbers, Can something special be recognized about these sums? One only learns recognizing special properties of numbers by participating in a mathematical discourse, a critical foundation of one's learning experience. In the context of this chapter, our experience includes the sums of two consecutive natural numbers. We not only have observed that such sums are not the multiples of two, but articulated this fact in a formal language of mathematics as well. So, one can look at the sum of three consecutive natural numbers from the multiplicative perspective; that is, to connect the concept of addition to that of multiplication by finding the average of three consecutive natural numbers. One can check to see that all such sums found above, namely, the numbers 6, 9, 12, 15, and 18 are multiples of three. How can this purely numeric observation be explained in terms of the first-order symbols, e.g., through the use of tape diagrams? The term "tape diagram" can be found in the Common Core State Standards [2010] with reference to a "drawing that looks like a segment of tape, used to illustrate number relationships" (p. 87). This term can be extended to any drawing/representation created out of square tiles (counters). Fig. 6.8 (far left) shows how the sum $1 + 2 + 3$ represented through square tiles can be put in three equal groups by finding the average number of the tiles among the elements of the group. This only requires moving a single tile up, an action that yields a 3×2 rectangle. A move from the sum $1 + 2 + 3$ to the sum $2 + 3 + 4$ requires the augmentation of each element by a single tile. In general, one can denote the first number in the sum of three consecutive natural numbers as n so that it will be followed by $n + 1$ and by $n + 2$ in the sequel. The resulting sum is $n + (n + 1) + (n + 2) = 3n + 3 = 3(n + 1)$; that is, the sum represents $n + 1$ tiles repeated three times. In other words, the sum is three times as big as the second number in the sum; alternatively, the second number is the average of the three numbers. This is how one can introduce the notion of average showing its connection to the sum of numbers the average value of which is sought and then dividing this sum by the number of addends. On the action level, the averaging always requires a single (invariant) action – moving the left-bottom square tile up to fill the top-left corner.

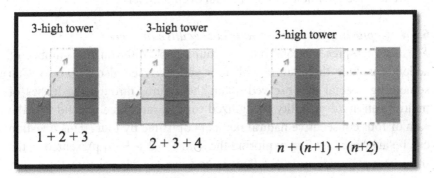

Fig. 6.8. The sum of three consecutive integers is a multiple of three.

As an illustration of interplay between procedural and conceptual knowledge, consider the following problem:

> *The sum of three consecutive natural numbers is equal to 81.*
> *Find the numbers.*

This problem seeks information about certain type of integers three of which have a given sum. Its procedurally informed algebraic solution is based on understanding that any three consecutive integers form an arithmetic progression with difference one. Thus, the three integers can be written as n, $n + 1$, and $n + 2$ from where the equation $n + (n + 1) + (n + 2) = 81$, then the value $n = 26$, and, finally, the triple of integers $(26, 27, 28)$ follow.

The problem, however, can be solved differently, in a purely arithmetical way, without any explicit use of algebra. Instead, by possessing conceptual understanding of the problem structure one can divide 3 into 81 to get 27 which then has to be diminished and augmented by one yielding 26 and 28, respectively. Similarly, one can find three consecutive integers with difference two and sum 81, only in that case, 27 has to be diminished and augmented by two yielding 25 and 29, respectively. A useful task is to show using a tape diagram that whereas the sum of any three consecutive integers with the given difference is divisible by three, the very sum may not always be an odd number. For example, 9, 14, and 19 are three consecutive integers with difference five and their sum is 42 – an even number.

6.5.4 *Exploring the sums of four consecutive integers*

The above representations put in groups depending on the number of addends include three sums with four addends (see also Fig. 6.5). Can something special be observed about these sums through the lenses of multiplicativity/divisibility recognized for two and three addends? Is the sum of four consecutive natural numbers divisible by four? This question can be answered by first exploring the sum $1 + 2 + 3 + 4$ presented in the form of a staircase in Fig. 6.9. It is because ten tiles cannot form four towers of equal height (alternatively, a 4-step staircase cannot be rearranged into a rectangle with side four), augmenting them by four, eight, twelve, and so on tiles; that is, by any quantity being a multiple of four, does not allow for the construction of four towers all the same height (forming a rectangle with side four). In an algebraic form, the sequence $x_k = 10 + 4k$ does not generate multiples of four, whatever the (integer) value of k.

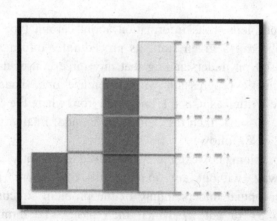

Fig. 6.9. The 4-step staircase cannot be rearranged into a rectangle with side four.

Remark 6.1. One can come to the following conclusion: if object A possess property P and object B does not possess property P, then jointly, A and B do not possess property P. For example, $4k$ is a multiple of four and 10 is not a multiple of four – therefore, $10 + 4k$ is not a multiple of four. It is helpful to ask teacher candidates if they could offer examples to support this conclusion. One such example is that the sum of an even and an odd number is not even but odd. That is, if P is the property of being even, then the sum does not possess this property. However, if P is the property of being odd, then the sum does possess this property. Likewise, the product of two numbers of different parity is an even number. That is, if a is even (property P) and b is odd, then the product ab is even (property P). However, if a is odd (property P) and b is even, then ab is even. So, the accuracy of the above conclusion depends on the nature of property P and an arithmetical operation involved.

Remark 6.2. Several other questions can be asked at that point. Can a square be made out of two squares? Is it always possible? Can this be done in more than one way? Consider the sum of two squares. Consider the product of two squares. How can one show by using concrete materials (e.g., square titles) that the sum of two squares may or may not be a square? How can one show by using square tiles that the product of two squares is always a square? Can the difference of two squares be also a square? Why or why not? How can this be connected to what was demonstrated (asked) previously? Fig. 6.10 shows how a 3×3 square is repeated four times yielding a 6×6 square; one can see that the product of two squares is a repetition of a square by a square number of times. (The next square will be a result of repeating the dark square nine times). Fig. 6.11 shows that adding a 3×3 square and a 4×4 square yields a 5×5 square; in other words, the sum of two squares may sometimes (but not always) be a square. This is a famous problem from geometry and number theory associated with the name of Pythagoras[14].

---

[14] Pythagoras (c. 570 BC – c. 495 BC) – a Greek mathematician and philosopher. The following quote translated from Russian into German and then into English can be found in Cooke [2010]: "Fifteen hundred years before the time of Pythagoras ... the Egyptians constructed right angles by so placing three pegs that a rope measured off into 3, 4, and 5 units would just reach around them, and for this purpose professional "rope fasteners" were employed" (p. 471).

Fig. 6.10. A square, 6^2, as a product of two squares: $2^2 \cdot 3^2 \ (= 3^2 + 3^2, + 3^2 + 3^2)$.

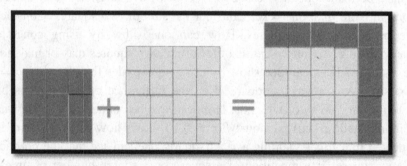

Fig. 6.11. A square, 5^2, as a sum of two squares, $3^2 + 4^2$.

Finally, our sums included two sums with five addends: 15 and 20 (see also Fig. 6.6). They both are multiples of five. The above observations can lead to the following

Conjecture about Trapezoidal Numbers

A sum of an even number of consecutive natural numbers is not a multiple of that number; a sum of an odd number of consecutive natural numbers is a multiple of that number. Put another way, a trapezoidal number is a multiple of the number of rows in its representation only when this number is odd.

6.6 Proof of the Conjecture about Trapezoidal Numbers

A physical representation of the proof of this conjecture for $n = 2, 3$, and 4 (Figs. 6.4–6.6) can inform its formal proof. That is, the first-order symbols provide foundation for using the second-order symbolism associated with proof. Indeed, without loss of generality, it is suffice to consider the sums that start with the number 1. This typical (for a professional mathematician) reduction to a special case, allowing one to consider it as a general case (that is, when a sum starts with the number 1), can be explained using tape diagrams as support system. Indeed, the sum of $2n - 1$ or $2n$ consecutive natural numbers that starts with a, after each a is split into $a - 1$ and 1, can be replaced by two sums: the first sum has either $2n - 1$ or $2n$ numbers equal to $a - 1$; the second sum consists of the first $2n - 1$ or $2n$ consecutive natural numbers. This is shown for $a = 3$, and $n = 3$ in Fig. 6.12 and for $a = 4$ and $n = 4$ in Fig. 6.13. Symbolically, we have either

$$a+(a+1)+(a+2)+...+(a+2n-2)$$
$$=[(a-1)+1]+[(a-1)+1+1]+[(a-1)+1+2]+...$$
$$+[(a-1)+1+2n-2]=(2n-1)(a-1)+(1+2+3+...+2n-1)$$

or

$$a+(a+1)+(a+2)+...+(a+2n-1)$$
$$=[(a-1)+1]+[(a-1)+1+1]+[(a-1)+1+2]+...$$
$$+[(a-1)+1+2n-1]=2n(a-1)+(1+2+3+...+2n).$$

The first addend in each of the two sums is either a multiple of $2n - 1$ or $2n$; the second addend does not have this obvious characterization. So, in order for the whole sum to be a multiple of either $2n - 1$ or $2n$, the sum of the first $2n -1$ or $2n$ natural numbers has to be a multiple of either $2n - 1$ or $2n$, respectively ($n = 2$ in Fig. 6.12 and Fig. 6.13). Therefore, considering the sums of the first $2n - 1$ or the first $2n$ consecutive natural numbers instead of a sum that starts with a number different from one is flawlessly general.

Using formula (6.1) yields

$$1+2+3+...+2n = 2n(2n+1)/2 = n(2n+1)$$

and

$$1+2+3+...+2n-1=(2n-1)2n/2=n(2n-1).$$

The former sum has $2n$ terms (that is, an even number of terms) and the product $n(2n + 1)$ is not divisible by $2n$ (after cancelling n out we see that $2n + 1$ is not divisible by 2). The latter sum has $2n - 1$ terms (that is, an odd number of terms) and the product $n(2n - 1)$ is obviously divisible by $2n - 1$. This completes the proof of the conjecture about trapezoidal numbers.

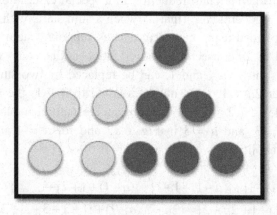

Fig. 6.12. Divisibility by three depends on the sum $1 + 2 + 3$.

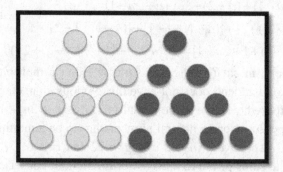

Fig. 6.13. Divisibility by four depends on the sum $1 + 2 + 3 + 4$.

6.7 Sums in Pairs of Odds and Evens

Analyzing the sums put in groups according to the first addend (section 6.4.2), that is, the sums that start with 1, 2, 3, and so on, one can see that in each of the groups the sums go in pairs – two even sums are followed by two odd sums and vice versa. This phenomenon can first be observed for triangular numbers – 1, 3, 6, 10, 15, 21, 28, 36, ... – which are the sums of consecutive natural numbers starting from one. It then continues for trapezoidal numbers, whatever the top row of a trapezoid is. For example, as was shown above for the sums that start with 2, we have

2 + 3 = 5, 2 + 3 + 4 = 9, 2 + 3 + 4 + 5 = 14, 2 + 3 + 4 + 5 + 6 = 20;

that is, the odd pair (5, 9) is followed by the even pair (14, 20). Also, whatever the rank of a triangular number is, through the transition to the sums that start with 2, its value is increased by that rank. For example, the sum 1 + 2 + 3 + 4 = 10 (the triangular number of rank four), the sum 2 + 3 + 4 + 5 = 14 (a four-row trapezoid the top row of which is two), and the difference between the two sums is four (14 – 10 = 4). On a physical level (Fig. 6.14), in order to rearrange a triangle built out of 10 counters into a trapezoid the top and the bottom rows of which are 2 and 5, respectively, one removes the top counter from the triangle and adds to it four counters to form the five-counter base. In general, the difference between the sum 2 + 3 + ... + n + 1 and 1 + 2 + ... + n is equal to n. These numerically supported observations and their generalization in the form of the relation $t_{n+1} - t_n = n$ (see formula (6.2) above) can motivate new algebraic explorations.

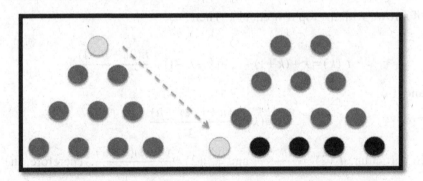

Fig. 6.14. Turning triangle into trapezoid requires four new counters.

6.7.1 *Learning to generalize from special cases*

To begin, consider two consecutive triangular numbers $t_n = \dfrac{n(n+1)}{2}$ and $t_{n+1} = \dfrac{(n+1)(n+2)}{2}$ of ranks n and n + 1, respectively. Then, because

$$t_n + n = \frac{n(n+1)}{2} + n = \frac{n(n+3)}{2} \ ,$$

it follows that

$$t_{n+1} + n + 1 = \frac{(n+1)[(n+1)+3]}{2} = \frac{(n+1)(n+4)}{2}.$$

Furthermore, $t_n + n = 2 + 3 + \ldots + (n+1)$. The last relation for $n = 3$ is shown in Fig. 6.15.

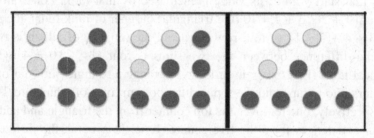

Fig. 6.15. $t_3 + 3 = \frac{3 \cdot 6}{2} = 2 + 3 + 4$.

Let $t_n(k)$ represent the sum of n consecutive natural numbers starting with k; in other words, $t_n(k)$ is a trapezoidal number comprised of n rows with the top row equal k. Then

$$t_n(k) = k + (k+1) + \ldots + (k+n-1) = \frac{n(2k+n-1)}{2}$$

and

$$t_{n+1}(k) = \frac{(n+1)(2k+n)}{2}.$$

In particular, $t_n(2) = \frac{n(n+3)}{2}$ and $t_{n+1}(2) = \frac{(n+1)(n+4)}{2}$. Therefore, the transition from the pair (t_n, t_{n+1}) of triangular numbers of the ranks n and $n + 1$ to the pair $(t_n(2), t_{n+1}(2))$ of trapezoidal numbers of the same ranks is as follows:

$$\left[\frac{n(n+1)}{2}, \frac{(n+1)(n+2)}{2} \right] \to \left[\frac{n(n+3)}{2}, \frac{(n+1)(n+4)}{2} \right]. \qquad (6.4)$$

Note that the difference between the second element of the first pair and the first element of the second pair in (6.4) is equal to one. Indeed,

$$\frac{(n+1)(n+2)}{2} - \frac{n(n+3)}{2} = \frac{n^2+3n+2-n^2-3n}{2} = 1.$$

This is shown in Fig. 6.16 through the first-order symbols for $n = 2$. In general (in terms of n and k),

$$t_{n+1} - t_n(k) = \frac{(n+1)(n+2)}{2} - \frac{n(2k+n-1)}{2}$$

$$= \frac{n^2+3n+2-2kn-n^2+n}{2} = 2n - kn + 1. \qquad (6.5)$$

Setting $k = 2$ in the far right term in (6.5) yields one.

At the same time, the difference between the second element of the second pair and the first element of the first pair in (6.4) is twice the rank of the second element of either pair (in Fig. 6.16 where $n = 2$ this difference is six). In general (in terms of n and k),

$$t_{n+1}(k) - t_n = \frac{(n+1)(2k+n)}{2} - \frac{n(n+1)}{2}$$

$$= \frac{n^2+2kn+n+2k-n^2-n}{2} = k(n+1). \qquad (6.6)$$

Setting $k = 2$ in the far right term in (6.6) yields $2(n+1)$.

Using the first-order symbols in demonstrating the meaning of algebraic relationships (depending on a single variable) in special cases (of this variable) facilitates the process of generalizing from these relationships further by introducing another variable. Indeed, a trapezoid described by the number $t_n(k)$ is defined by its top row k and the number of rows n. So, the first-order symbols play important role in supporting transition to the second-order symbolism and its various generalizations.

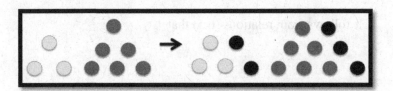

Fig. 6.16. Relations (6.4) – (6.6) in the case $n = 2$.

6.7.2   *Comparing triangles to trapezoids with the top row greater than two*

The next observation to be explained is why even and odd sums go in pairs when consecutive natural numbers are added. To this end, let the sum of n consecutive natural numbers (n is an even number), starting from an even number k be an even number $2m$; that is, let

$$k + (k + 1) + (k + 2) + \ldots + (k + n - 1) = 2m.$$

Then, augmenting this sum by an even number $k + n$ yields $2m + k + n$, an even number, which turns into an odd number when augmented by the next (odd) number $k + n + 1$. The next augmentation by an even number, $k + n + 2$, makes the sum an odd number which turns into an even number as it is augmented by an odd number $k + n + 3$. This cycle will continue. The case $k = 2$, $n = 4$, and $2m = 2 \cdot 4 + \underbrace{(1 + 2 + 3)}_{even}$ is shown in Fig. 6.17. As an exercise, other combinations of k and n may be considered.

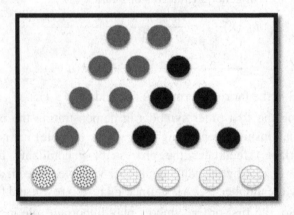

Fig. 6.17. A sum with all even components:
$2 \cdot 4 + (1 + 2 + 3) + (2 + 4) = 2 + 3 + 4 + 5 + 6 = 20$.

It follows from relations (6.5) that

$$t_n(k) = t_{n+1} - (2n - kn + 1) = \frac{(n+1)(n+2)}{2} - 2n + kn - 1$$

$$= \frac{n^2 + 3n + 2 - 4n + 2kn - 2}{2} = \frac{n(n+2k-1)}{2}.$$

That is,

$$t_n(k) = \frac{n(n+2k-1)}{2}. \qquad (6.7)$$

It follows from relations (6.6) that

$$t_{n+1}(k) = t_n + k(n+1) = \frac{n(n+1)}{2} + k(n+1) = \frac{(n+1)(n+2k)}{2}.$$

That is,

$$t_{n+1}(k) = \frac{(n+1)(n+2k)}{2}. \qquad (6.8)$$

Furthermore, it follows from (6.7) and (6.8), respectively, that

$$t_n(k) - t_n = \frac{n(n+2k-1)}{2} - \frac{n(n+1)}{2} = \frac{n^2 + 2kn - n - n^2 - n}{2}$$

$$= n(k-1)$$

and

$$t_{n+1}(k) - t_{n+1} = \frac{(n+1)(n+2k)}{2} - \frac{(n+1)(n+2)}{2} = (n+1)(k-1).$$

That is,

$$t_n(k) = t_n + n(k-1). \qquad (6.9)$$

Formula (6.9) shows that augmenting a triangular number by the $(k-1)$-multiple of its rank yields the trapezoidal number $t_n(k)$. The case $k = 2$ and $n = 3$ is shown in Fig. 6.15. One can be asked to show that formula (6.9) is a rather obvious relation demonstrating how to transform the sum of the first n natural numbers into the sum of n consecutive natural numbers starting with the number k. Making this connection allows one to see how mathematical relations that require rather complex symbolic manipulation have in fact a rather obvious meaning. This is why mathematical connections typically have strong conceptual power and therefore have been emphasized throughout various standards and recommendations for the teaching of mathematics around the world.

6.8 Mathematical Knowledge Used for Teaching Young Children

Another value of the Basic Task discourse in a mathematics teacher education classroom is that starting from a hands-on activity and extending it to formal mathematical knowledge, a teacher candidate can use this knowledge to design new grade-appropriate hands-on activities for young children who will be engaged into conceptually rich mathematical tasks without knowing anything about their richness. For example, knowing that number 15 has three representations as the sums of consecutive natural numbers:

$$15 = 1 + 2 + 3 + 4 + 5, \ 15 = 4 + 5 + 6, \text{ and } 15 = 7 + 8,$$

the following hands-on activity can be offered. Give young children 15 counters and ask them to build different trapezoids under the condition that each row of a trapezoid has one more/fewer counters that the one immediately above/below it. This kind of task may include 16 counters with the goal to show that trapezoids (let alone triangles) cannot always be constructed. This experiment can continue with 8 and then 4 counters to the same effect. Then, one can be asked what is special about the numbers 2, 4, 8, and 16? This question may also be given a hands-on format – how can one get four counters out of two, how can one get eight counters out of four, 16 counters out of eight, and so on? The idea of doubling may be easier developed at the level of the first-order symbols.

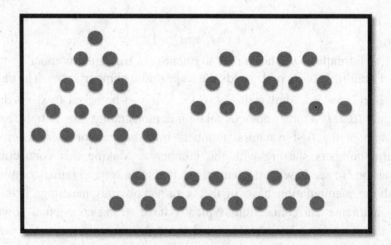

Fig. 6.18. Representing 15 as a trapezoidal number in three ways.

Alternatively, one can use formula (6.3) which, as was mentioned at the conclusion of section 6.3, suggests that if an integer's double cannot be represented as a product of two factors of different parity, the integer itself cannot be represented as a sum of consecutive natural numbers; in other words, it does not have a trapezoidal representation. This explains in the formal language of mathematics why the powers of two could not be found among the sums of consecutive natural numbers: their doubles always remain powers of two and thus do not have odd factors.

6.9 How Many Trapezoidal Representations Does a Number Have and How Can One Find Them?

How many trapezoidal representations does each number in the range [3, 20] have? First of all, note that if a number is odd, then it has at least one representation through the sum of two consecutive natural numbers. However, some odd numbers, like 9 and 15, have more than one trapezoidal representation. Are there even numbers with more than one such representation? As was shown in section 6.4, the only such number in the range [3, 20] is 18. Indeed, $18 = 5 + 6 + 7$ and $18 = 3 + 4 + 5 + 6$. To answer these questions, consider one more time formula (6.3) from section 6.3.3:

$$k+(k+1)+...+n = \frac{(n+k)(n-k+1)}{2}.$$

The formula is much more informative than it could appear from the first glance. Yet, its proper analysis is rather simple. Noting that $n + k > n - k + 1$ for all natural numbers n and k (indeed, as was shown in section 6.3.3, the difference $n + k - (n - k + 1) = 2k - 1 > 0$ for $k = 1, 2, 3, ...$), we have to factor twice the product of a candidate for a trapezoidal representation in two unequal factors of *different* parity, $n + k$ and $n - k + 1$, where the smaller factor is equal to the number of addends in the sum (alternatively, the number of rows in a trapezoid). For example, one such representation for the number 18 is as follows: $18 = \frac{36}{2} = \frac{12 \cdot 3}{2}$. This tells us that there is a trapezoid with three rows. Therefore, 18 can be represented as a sum of three consecutive natural numbers with the number 6 (=18÷3) being the average of the three numbers: $18 = 5 + 6 + 7$. Also, $18 = \dfrac{9 \cdot 4}{2}$ indicating that a trapezoid with four rows can be created out of 18 counters. In order to find the first and the last terms of the sum sought, one has to find two numbers, n and k, $n > k$, which sum is 9 and difference 3. These numbers are $n = 6$ and $k = 3$. That is, $18 = 3 + 4 + 5 + 6$.

Is there another way to represent 18 as half the product of two integers of different parity? The only other option is $18 = \dfrac{36 \cdot 1}{2}$, but in that case the number of summands (rows in a trapezoid) is equal to one, a trivial representation of 18 through itself. Thus, 18 has only two (non-trivial) trapezoidal representations and they may be counted by the number of odd factors of 18 that are different from one.

Let us see how this reasoning works in the case of the number 15 for which we already found three representations. We have, $15 = \dfrac{15 \cdot 2}{2} = \dfrac{10 \cdot 3}{2} = \dfrac{6 \cdot 5}{2}$. The first product points at the representation with two addends: $15 = 7 + 8$. The second product points at the representation with three addends of which the average addend is equal to 5; that is, $15 = 4 + 5 + 6$. Finally, the third product points at two things: there is a representation with five addends and it is a triangular representation (as the factors 5 and 6 differ by one); that is, $15 = 1 + 2 + 3 + 4 + 5$. Once again, we have three products, $15 \cdot 2, 10 \cdot 3$, and $6 \cdot 5$, each of which included an odd factor of 15 greater than one. These factors (namely, 15, 5, and 3) can be used as counters of the number of trapezoidal representations of the number 15.

Remark 6.3. Let m and n, $m > n$, be natural numbers of different parity and $N = \dfrac{mn}{2}$. Then N has a trapezoidal representation with n rows.

Remark 6.4. If N is a prime number, then it has exactly one trapezoidal representation. Indeed, the only way to factor $2N$ in two factors different from the unity is $N \cdot 2$. This factoring points at a trapezoid with two rows; that is, $N = \dfrac{N-1}{2} + \dfrac{N+1}{2}$. For example, when $N = 13$ we have $\dfrac{N-1}{2} = 6$ and $\dfrac{N+1}{2} = 7$.

Remark 6.5. The product of a prime number and a power of two has only one trapezoidal representation. The number of rows in the corresponding trapezoid is equal to either the prime number or twice the power of two depending on which of the two quantities is larger. For example, when $N = 52 \, (= 13 \cdot 4)$ we have the relation $52 = \dfrac{13 \cdot 8}{2}$ pointing at the representation with 8 rows. At the same time, when $N = 104$ we have the relation $104 = \dfrac{16 \cdot 13}{2}$ pointing at a trapezoid with 13 rows. It remains to find the length of the top (or the bottom) row in each case. In the case of 52 one has to find positive integers n and k, $n > k$, the sum of which is 13 and difference 7. We have $n = 10$ and $k = 3$. That is, $52 = 3 + 4 + 5 + 6 + 7 + 8 + 9 + 10$. Likewise, in the case of 104 one has to find positive integers n and k, $n > k$, the sum of which is 16 and difference 12. We have $n = 14$ and $k = 2$. That is, $104 = 2 + 3 + ... + 14$.

Remark 6.6. Once the first trapezoidal representation with 13 rows has been found, the values of k and n for all other representations of the product of 13 and a power of two (that is, for 208, 416, 832, and so on) can be found through the following recursive formulas $k_i = k_{i-1} + 2^{i+2}, n_i = n_{i-1} + 2^{i+2}, k_0 = 2, n_0 = 14, i = 1, 2, 3, ...$. For example, when $i = 3$ we have
$$k_3 = k_2 + 2^5 = k_1 + 2^4 + 2^5 = k_0 + 2^3 + 2^4 + 2^5 = 2 + 2^3 + 2^4 + 2^5 = 58,$$
and
$$n_3 = n_2 + 2^5 = n_1 + 2^4 + 2^5 = n_0 + 2^3 + 2^4 + 2^5 = 14 + 2^3 + 2^4 + 2^5 = 70.$$
Therefore, $104 \cdot 8 = 58 + 59 + 60 + ... + 70 = 832 = 13 \cdot 2^6$.

Remark 6.7. An interesting property of integers, the only factors of which are a power of two and a prime number, follows from Remark 6.3. If one starts with a prime number p, whatever its value, it has a single

trapezoidal representation with two rows due to the relation $p = \dfrac{p \cdot 2}{2}$.
Consequently, multiplying the prime number p by a power of two, after some transient period (during which we have trapezoids with the number of rows equal to the powers of two), the corresponding products $p \cdot 2^m$, $p < 2^{m+1}$, would have trapezoidal representations with the number of rows equal to the prime number p only.

Chapter 7

Informal Geometry

7.1 Introduction

One of the challenges facing mathematics education community in the digital era is the appropriate integration of technology into the elementary school mathematics curriculum and, in particular, the improvement of teaching and learning geometry at that level. The ease of computing provided by technology changes the way students think about mathematics, and it brings about opportunities for new content, new curricula, and new teaching methods. The appropriate use of software tools provides tremendous potential for advancing the ideas of mathematical experimentation in the classroom. This approach assumes the intertwining of teacher demonstration and student independent investigation, and it uses computer as a means for enhancing concept formation, developing mathematical reasoning, stimulating conjecturing, linking different mathematical ideas, and promoting the image of mathematics as an exploratory science.

Research carried out within the last three decades suggests that the ease of using computer software in the teaching of mathematics may have some undesirable consequences because it encourages students, including elementary teacher candidates, to use empirical data as a substitute for formal demonstration. In particular, there is evidence that dynamic geometry software such as *The Geometer's Sketchpad* [Jackiw, 1991], *CABRI Geometry* [Laborde, 1991], and *GeoGebra* [Hohenwarter, 2002] run this risk by evoking a faulty appreciation of their measuring facilities [Balacheff, 1988; Yerushalmy, 1990; Chazan, 1993; Kakihana et al., 1994; Hoyles et al., 1995; Marrades and Gutierrez, 2000; Hollebrands, 2007; Carreira et al., 2016]. For instance, in a frequently cited paper, Chazan [1993] presented an example in which students, after partitioning a triangle by a median in two parts are inclined to make a conjecture about equal areas of the parts by performing measurements on a series of triangles in support of the conjecture. As a result, they see no reason for a formal justification of the conjecture. This empiricist view of mathematics emerges because such use of measurement in the exploration requires one only to recognize the *sameness* that can easily be captured by comparing *two* numbers. A computational experiment involving the comparison of two quantities and encouraging making a general statement after a small number of measurements can develop inadequate reasoning based on unwarranted generalization and may prevent learners of mathematics from appreciating the use of mathematical symbolism as a means of formal deduction of a mathematical proposition. On the other hand, such an experiment can bring about a counter-example to an erroneous conjecture that the same (i.e., the congruency of two parts) is true for an altitude of a triangle and thus measurement can be utilized to defy the conjecture. That is, depending on context, the same computational method can be used either to support mathematical reasoning or to lead it astray. Whereas there is no definitive rule indicating the case of support, empirical induction as a means of generalization should be used in conjunction with deductive reasoning through which one can make sense of numerical patterns observed due to the ease of computing.

A numerical approach to geometry based on a computational experiment has ancient roots as evidenced by the following historically remarkable statement made by Archimedes[15] [1912] in a letter to Eratosthenes[16], "Certain things first became clear to me by a mechanical method, although they had to be demonstrated by geometry afterwards because their investigation by the said mechanical method did not furnish an actual demonstration. But it is of course easier, when the method has previously given us some knowledge of the questions, to supply the proof than it is to find it without any previous knowledge" (p. 13). In this quote one can recognize the importance of the duality of informal geometry and formal reasoning.

How can the power of computing technology be used to promote the idea of formal demonstration (proof) in mathematics teacher education and in school mathematics rather than further contribute to 'the death of proof'? How can this power be put to work to bridge empirical and deductive reasoning in exploratory learning context? If the purpose

---

[15] Archimedes (c. 287 BC – c. 212 BC) – a Greek mathematician, by some accounts (e.g., [Rohlin, 2013]) is the greatest mathematician of all time.

[16] Eratosthenes (c. 276 BC – c. 194 BC) – a Greek scholar.

of introducing elementary teacher candidates to simple mathematical proofs is to promote conceptual understanding [Hersh, 1993], then such understanding has to be supported and motivated through a meaningful instructional activity. In the words of Hanna [1995], 'one must structure and present the proof in such a way that is clear and convincing, and one must equip the students with the tools that will allow them to understand a proof' (p. 47). Debating the use of dynamic geometry software as an exploratory tool in an educational context, Christou *et al.* [2004], based on their experience working with elementary teacher candidates, argued that such kind of "exploration is not inconsistent with the view of mathematics as an analytic science or with the central role of proof" (p. 342). As far as educational uses of computers are concerned, this position calls for mathematics educators to design appropriate activities for students and their future teachers alike that could help them bridge informal computer-based explorations of grade-appropriate mathematical ideas and formal presentation of the ideas in a convincing way.

In this chapter, the joint use of *The Geometer's Sketchpad* and a spreadsheet with the goal to address the above concerns by reflecting on activities that structure formal presentation of mathematical ideas through a numerical approach will be discussed. In particular, a way of using the measuring facilities of the dynamic geometry program in an exploratory study of mathematics without coming into contradiction with rules and traditions of its practice built on formal presentation of mathematical concepts will be demonstrated. The material for this chapter was collected over the years in the context of the author's work with elementary teacher candidates enrolled in different mathematics education courses emphasizing the didactic value of informal geometry learning and the use of technology [Abramovich and Brown, 1999; Abramovich, 2010a, 2016].

7.2 Geoboard Explorations

A geoboard, introduced by Gattegno [1971] as a hands-on learning environment for exploring basic geometric ideas, represents, conceptually, a rectangular grid in the form of a periodic array of points in a plane (sometimes called a plane lattice). Physically, the environment allows for the construction of a variety of polygons by using rubber bands held by pegs (alternatively, lattice points). A polygon on a geoboard can be associated with the number of pegs that the rubber band touches. Another characteristic of the polygon is the number of pegs in its interior. Thus polygons may be compared through the number of pegs related to each of them. For example, in one case a rubber band touches six pegs and encloses three pegs; in another case – touches eight pegs and encloses two pegs. (Can these be triangles of equal areas?).

This association of shapes with pegs on a geoboard brings about counting in the context of a geoboard as one of the major problem-solving strategies furnishing geoboard geometry with an informal flavor, something that is especially important at the elementary level. Furthermore, because on a geoboard a linear unit is a side of a unit square the vertices of which are the four pegs closest to each other, one can find area of any shape by using a strategy shown in Fig. 7.1 – enclose the shape into a rectangle (square) and then subtract from its area the (easy to find) areas of extraneous triangles as they always have half of area of a rectangle which, in turn, comprises unit squares. This is due to the fact, often confirmed through informal geometry of paper and scissors, that a diagonal of a rectangle cuts it in two congruent triangles. In other words, addition and subtraction are two major arithmetical operations in the context of finding areas on a geoboard.

Even before geoboards became available, such informal approach to geometry was used by Wertheimer [1959] and his followers with children as young as five-year-old in finding areas of rectangles, parallelograms, isosceles triangles, and trapezoids. As noted by Luchins and Luchins [1970a], "Wertheimer's method of finding the areas of geometric figures by transforming them into others, particularly rectangles and triangles, is an ancient one. It has been used, for example, in problems given in an Egyptian papyrus roll [dated circa 1650 BC], found in 1858 by [a Scottish scholar and collector of antiques] Henry Rhind" (p. 44). According to Boyer [1968], "In transformations ... in which isosceles triangles and trapezoids are converted into rectangles, we see the beginning of the theory of congruence and of the idea of proof in geometry, but the Egyptians did not carry their work further" (p. 180). It is interesting to see ancient roots in what one might consider as 'modern' methods of teaching geometry in Japanese elementary schools when helping students to understand "that the idea for finding the area of a triangle using the equivalent-area transformation of the rectangle into a parallelogram by bisecting the height can be used to determine the area of special triangles and trapezoids" [Takahashi *et al.*, 2004, p. 263]. So, the notion of diversity of mathematics teaching methods can be extended not only to include their international dimension but also the contribution of ancient Egyptian civilization that goes back to several millennia in the history of human race and mathematics, in particular. In that way, what really can be referred to a 'modern' method within the worldwide practice of mathematics teaching methods is the use of technology in the very broad sense of this word.

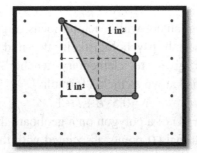

Fig. 7.1. Finding area of the shaded polygon using informal geometry.

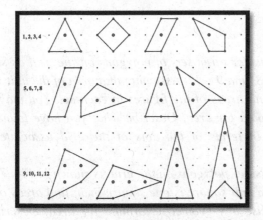

Fig. 7.2. Finding areas of the polygons
with $B = 4$ and $I=1,2, 3$.

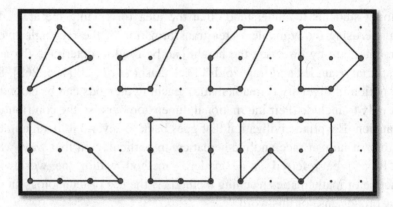

Fig. 7.3. Finding areas of the polygons with $I = 1$ and $B = 3, 4, ..., 9$.

So, in full agreement with teaching ideas originated in antiquity and practiced by Gestalt psychologists in the mid 20th century, the following task was given to elementary teacher candidates as an introduction to what is known as Pick's formula[17]

$$A = 0.5 \cdot B + I - 1 \qquad (7.1)$$

which expresses area (A) of a polygon on a geoboard through the number of border (B) and internal (I) points associated with the polygon. (As an aside, formula (7.1) confirms that the above-mentioned two polygons with pairs (B, I), respectively, (6, 3) and (8, 2) have indeed the same area).

Task 1.

a) Consider three sets of polygons on the grid as shown in Fig. 7.2: 1 – 4, 5 – 8, and 9 – 12. Assuming that area of the unit square is one square inch, find areas of the figures in each set. You may not use any geometric formula for area. Describe what you have found. Formulate your findings in terms of the dots of the grid associated with each polygon.

b) Consider polygons on the grid as shown in Fig. 7.3. Assuming that area of the unit square is one square inch, find areas of the figures. You may not use any geometric formula for area. Describe what you

---

[17] George Alexander Pick (1859–1942) – an Austrian mathematician.

have found. Formulate your findings in terms of the dots of the grid associated with each polygon.

The goal of Task 1 was to show empirically how by increasing the number of internal points by one increases area of the corresponding polygon by one square unit and how by increasing the number of border points by one increases area by half of the unit. This is a simplified approach to formula (7.1), through which the formula is introduced inductively as a way of confirming empirically found areas. Typically, without being guided by a 'more knowledgeable other', very few teacher candidates notice any dependence of area on the points (pegs) associated with the shape and a common response to their findings is that all polygons in set 1 – 4 have area two square units, in set 5 – 8 have area three square units, and in set 9 – 12 have area four square units. The teacher candidates were expected to find area of a polygon as shown in Fig. 7.1.

Finding areas of the shapes pictured in Fig. 7.2 by using the strategy shown in Fig. 7.1, easily yields the following triples of B, I, and A: $(B, I, A) \in \{(4, 1, 2), (4, 2, 3), (4, 3, 4)\}$. Likewise, finding areas of the shapes pictured in Fig. 7.3, easily yields the following triples of B, I, and A:
$$(B, I, A) \in \{(3, 1, 1.5), (4, 1, 2), (5, 1, 2.5), (6, 1, 3),$$
$$(7, 1, 3.5), (8, 1, 4), (9, 1, 4.5)\}.$$

Instructor-guided analysis of the former set of three triples typically results in the following observation: when the number of border pegs stays the same, with each additional internal peg, area increases by one square unit. Likewise, the analysis of the latter set of seven triples typically results in a similar observation: when the number of internal pegs stays the same, with each additional border peg, area increases by a half of the square unit. While teacher candidates can be guided to make these observations without any difficulty, conjecturing formula (7.1) had always been a true challenge. It appears that the major obstacle for using numerical evidence provided by the method of analyzing Fig. 7.2 and Fig. 7.3 has been due to the need for simultaneous coordination of two variables (B and I) involved, despite, or perhaps because of the linear structure of Pick's formula. Teacher candidates' immediate experience in finding areas of polygons through partitioning them into triangles and rectangles might have encouraged them to look for a product of variables; for example, some may offer in the case of Fig. 7.2 the formula $A = \dfrac{B}{4} \cdot I$. Indeed, when $B = 4$, the last formula confirms numerical evidence collected through exploring the shapes in Fig. 7.2. But already the case of the unit square where $I = 0$ provides a pretty convincing counter-example with $A = 0$. Therefore, a different set of tasks and the use of technology both as a generator and a processor of data are needed. The appropriate use of technology, in the (above-cited) words by Archimedes, will provide a mechanical method aimed to support conjecturing Pick's formula through a combination of inductive and deductive reasoning, two distinct but often complimentary methods of inquiry into mathematical structures.

Remark 7.1. In the collateral learning fashion, one can use the diagram of Fig. 7.4 in order to express areas of triangles and trapezoids through fractions. Doing that by using measuring facilities of *The Geometer's Sketchpad* may be a challenge because the program returns measurements in the form of a decimal fraction (Chapter 4) and therefore it will not be possible to measure exactly area equal 1/6. An alternative way is to demonstrate that the shaded triangle, cut off from the top unit square, by a diagonal of a rectangle that comprises three unit squares is 1/6 of the unit square. This can be demonstrated through rotation, another useful feature of the program.

Remark 7.2. The diagram of Fig. 7.4 can also be used to demonstrate the connection of basic geometric principles to Gestalt psychology, the essence of which Wertheimer [1938] formulated in the following way: "There are wholes, the behavior of which is not determined by that of their individual elements, but where the part-processes are themselves determined by the intrinsic nature of the whole" (p. 2). Indeed, in Fig. 7.4 the large shaded triangle may be considered as the whole; however, it is not defined by its three parts – one triangle and two trapezoids. Rather, the parts are defined by the whole triangle using the similarity of triangles: because the ratio of the legs of the whole is 3 to 1, so is the ratio of the legs of the small triangle (a part of the whole). This yields 1/3 as the length of the horizontal leg of the small triangle thus demonstrating that the triangular part of the whole is 1/6 of the unit square. From here it follows that the smaller and the larger trapezoidal parts of the whole are, respectively, 1/2 and 5/6 of the unit square. It is not surprising that, as was mentioned above, Gestalt psychologists used problem-solving strategies of informal geometry in teaching mathematics to young children grounding their pedagogy in the tenet, "What holds for the mathematical formula applies also to the "formula" of Gestalt theory" [*ibid*, p. 3].

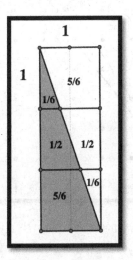

Fig. 7.4. Three unit squares – one diagonal.

Remark 7.3. A geoboard can be used to identify points the coordinates of which are relatively prime numbers; that is, natural numbers, the greatest common divisor of which is equal to one. Put another way, the coordinates of such points form irreducible fractions. Indeed, by connecting a point on the geoboard and the origin with a segment two cases can be considered: the segment passes and does not pass through another point of the geoboard (Fig. 7.5). In the latter case, the point selected has relatively prime coordinates, which form an irreducible fraction when used as the numerator and denominator; in the former case, the coordinates of the point form a fraction that can be reduced to the simplest form. As an example, consider the point (4, 5). The segment that connects this point with the origin does not pass through any other point of the geoboard and the fraction 4/5 is irreducible. At the same time, the segment connecting the origin with the point (4, 6) passes through the point (2, 3) and 4/6 = 2/3. This situation will be explored further in Chapter 8.

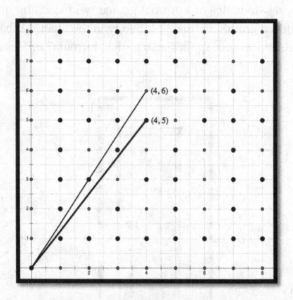

Fig. 7.5. Identifying points with relatively prime coordinates.

7.3 Towards a Double-Application Environment

In what follows, an alternative approach to Pick's formula that was used by the author in a technology-enhanced *General Mathematics* course for prospective teachers is described. This approach, in full agreement with the ideas of Wertheimer [1959] and his disciples piloted in their experimental work with young children in a mathematics classroom (and even in agreement with Egyptian ideas known as early as two millennia BC), allowed for the derivation of formula (7.1), credited to the modern era mathematician Pick, through a numerical approach. It was supported by the capability of *The Geometer's Sketchpad* to measure areas of polygons on a (computational) geoboard and enhanced by a spreadsheet's facility of recurrent counting.

The idea of using a spreadsheet, conceptualized educationally as "an electronic blackboard and electronic chalk in a classroom" [Power, 2000], stemmed not only from a belief that mathematical visualization strongly affects inductive reasoning. Whereas area on a geoboard varies recursively, a spreadsheet has a powerful feature of computing through recursive formula. The use of a spreadsheet makes it possible to enter experimental data into an electronic chart using another mechanical method involving cell referencing. Such a method is likely to promote, if not to trigger, the idea that area on a geoboard can be defined recursively. It is a recursive definition of area that has to be the outcome of empirical induction rather than a closed formula, namely, (7.1). Therefore, the role of a spreadsheet is to aid teacher candidates in the inductive discovery of a recursive formulation of the area and to encourage them to derive formula (7.1) from their discovery; in other words, to "use such a discovery as an opportunity to investigate more exactly the properties discovered" [Euler, cited in Pólya, 1954, p. 3].

The activities are composed of three major parts focusing on teacher candidates' independent investigation and whole class discussion mediated by instructor's demonstration. These include 1) using computational geoboards for finding areas of different polygons, entering the results of explorations (data) into spreadsheet templates, and finding numerical patterns within the templates, 2) demonstrating how one can use a spreadsheet for formal deduction of formula (7.1), and 3) using jointly *The Geometer's Sketchpad* and a spreadsheet for a formal deduction of formula (7.1), i.e., Pick's formula. The main didactic idea that permeates the activities is to help teacher candidates view the results of measurement not as a replacement of formal demonstration in geometry but rather as an important link in a chain of diverse computational experiments deductively leading to a conjecture about area on a geoboard and its articulation using the formal language of mathematics. These activities provide true opportunities for elementary teacher candidates "to conduct an investigation ... to engage in the use of a variety of technological tools ... to explore and deepen their understanding of mathematics, even if these tools are not the same ones they will eventually use with children" [Conference Board of the Mathematical Sciences, 2012, p. 34]. A variety of tools used in the context of a mathematical investigation at a higher level are, nonetheless, helpful for the appreciation of and competence in the diversity of methods that are appropriate to be used with children who, likewise, should be "able to use technological tools to explore and deepen their understanding of concepts" [Common Core State Standards, 2010, p. 7].

7.4 Guided Exploration on a Computational Geoboard

After having been introduced to the above-mentioned computational tools, teacher candidates were assigned the following two tasks to be carried out on a (computational) geoboard.

Task 2. *How does the area change if one starts from the fundamental triangle (B=3, I=0, Area = 0.5 square units), keeps B constant, and subsequently increases I by one? Describe the polygons you have constructed.*

Task 3. *How does the area change if one starts from the fundamental triangle (B=3, I=0, Area = 0.5 square units), keeps I constant, and subsequently increases B by one? Describe the polygons you have constructed.*

The completion of Tasks 2 and 3 by a teacher candidate are shown in Fig. 7.6 and Fig. 7.7, respectively. The teacher candidate explained the exploration pictured in Fig. 7.6 as follows.

> *The area of a figure, beginning with a fundamental triangle (B = 3, I = 0) and increasing I by 1 with B constant for each consecutive figure, increases by 1 inch [sic]. It is interesting to note that, when keeping two border points in the same place on the geoboard axis, the third border points of each consecutive figure fall along a line with slope of 2. That is, the third border point moves up two and over one increment on the axis with each new point drawn.*

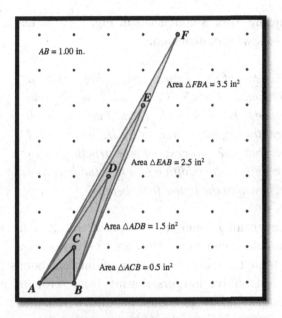

Fig. 7.6. Exploring Task 2.

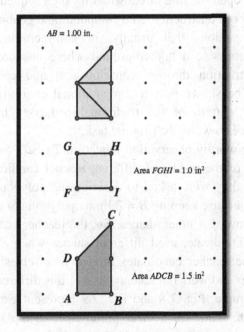

Fig. 7.7. Exploring Task 3.

Regarding Task 3 completed in Fig. 7.7, the teacher candidate provided the following explanation.

> *The area of a figure, beginning with a fundamental triangle (B = 3, I = 0) and increasing B by 1 with I constant for each consecutive figure, increases by 1/2 inch [sic]. I noted that the process of constructing each figure reflects the addition of a new fundamental triangle within the figure from before.*

An interesting, and unanticipated result that the teacher candidate was able to come up with when working on Task 2 is a rule for constructing triangles with the number of internal points forming the sequence of consecutive integers starting from zero. Indeed, it is not immediately apparent how to choose the third vertex of a triangle when adding a new interior point without altering the number of border points. Task 2 leaves it open the rule of constructing such sequence of triangles. Keeping records of construction on a computational geoboard allows one to move from actions that usually, in non-recording media, lack conscious awareness to a higher ground where one could master the geometric construction through consciousness and control. In what follows, it will be shown how a computational environment alters the question-answer pattern of the traditional pedagogy by encouraging reflection that goes beyond the original task.

Another worthy observation related to Tasks 2 and 3 deals with the comparison of the results of different teacher candidates. Note that the very reason they were asked to describe the polygons used in their explorations is because keeping $B = 3$ limits polygons to triangles, while increasing B allows for other shapes. So, the teacher candidate, as Fig. 7.6 and Fig. 7.7 indicate, used different shapes when exploring areas. Yet, most of the teacher candidates limited themselves to triangles in both explorations and were not able to observe this difference.

For example, Fig. 7.8 and Fig. 7.9 represent explorations by a teacher candidate who had a clear preference, as the wording of his comments indicates, to think of polygons in terms of triangles only. A reason for such a preference might be that the first exploration was geometrically limited to triangles and this influenced his thinking in the subsequent task. Analyzing work of other teacher candidates showed that those who constructed different shapes used a more general word 'figure' when describing their construction of triangles, whereas the teacher candidate who demonstrated a predilection for triangles in both constructions used more specific terms, like triangles and quadrilaterals.

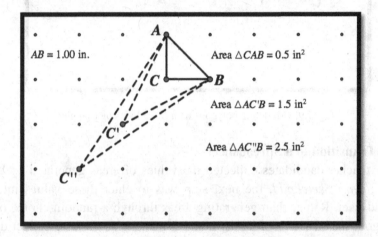

Fig. 7.8. Emphasis on triangles in wording.

Remark 7.4. The observed preference for constructing triangles may also be explained in terms of Einstellung effect [Luchins, 1942] – a psychological tendency to use successful problem-solving strategy of one's past experience in situations that either can be resolved more efficiently or to which the strategy is not applicable at all. Luchins and Luchins [1994] noted that Einstellung effect is not necessarily a deterrent to productive thinking – one of the pillars of Gestalt theory – but rather a phenomenon when past experience creates conditions for the rigidity of thinking (making it mainly reproductive), something that has a negative impact on creativity. In particular, the rigidity of thinking might lead to an incorrect generalization like the one made by a teacher candidate regarding all isosceles triangles with one internal point having an even number of border points on a geoboard.

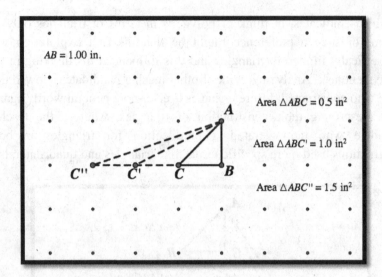

Fig. 7.9. Modeling polygons with $B > 3$ by using triangles.

7.5 Transition to a Spreadsheet

After teacher candidates collected the values of areas generated by *The Geometer's Sketchpad*, the next step was to enter these values into a spreadsheet. Rather than generating areas through a random choice of a polygon, the structure of Tasks 2 and 3 made it possible to collect data in a systematic way, agreeable with the geometric structure of the spreadsheet template. In doing so, teacher candidates could comprehend the concept of boundary conditions in informal way. In other words, the pedagogy of informal geometry is conducive to the development of a concept through an action with deferred appreciation of mathematical significance of both the concept and the action.

When the boundary conditions were set up on a spreadsheet, teacher candidates were asked to switch their computer screens back to *The Geometer's Sketchpad* and to explore the case when starting from the fundamental triangle both B and I increase by one. Rather than thinking of the case $B = 4$, $I = 1$ in terms of quadrilaterals (e.g., see the quadrilateral in Fig. 7.1), most of the teacher candidates once again, demonstrating Einstellung effect, limited themselves to triangles when exploring this special case (e.g., see the triangle in the top left corner of

Fig. 7.2). That is, the preference for triangles was steadily demonstrated and parallelograms as possible quadrilaterals with one internal point were avoided. One can suggest that such mind set, however, is deeply rooted in an inborn character of spatial thinking which has clear preference for 'upright' rather than for 'oblique' with a fixed link to the vertical axis in one's cognitive space [DeSoto *et al.*, 1965; Knauff and May, 2006]. Indeed, even the fundamental triangles (i.e., $B = 3$, $I = 0$) that teacher candidates have constructed are mainly right triangles.

In this respect, another 'result' that one teacher candidate produced through exploring the case $B = 4$, $I = 1$ is worth noting. In modeling this special case, she used an isosceles triangle with a vertical line of symmetry and then went on to claim that on a geoboard "*all* isosceles triangles with one internal point have even number of border points". It might have been the predilection for 'upright' (another case of Einstellung effect) turned her eyes away from an isosceles triangle with an oblique axis of symmetry (see triangle ABC' in Fig. 7.8) thus yielding the above unwarranted generalization.

So, despite the ease of measuring areas, teacher candidates felt more comfortable with constructing a triangle with $B = 4$ and $I = 1$. Measuring its area yielded the number 2 and this value was entered into the appropriate cell of a spreadsheet. Fig. 7.10 shows the boundary conditions for area (B2:H2 and B2:B8) and the number 2 displayed in cell C3.

C3		▼		2			
	A	**B**	**C**	**D**	**E**	**F**	**G**	**H**
1	B\I	0	1	2	3	4	5	6
2	3	0.5	1.5	2.5	3.5	4.5	5.5	6.5
3	4	1	2					
4	5	1.5						
5	6	2						
6	7	2.5						
7	8	3						
8	9	3.5						

Fig. 7.10. Approaching the case $B = 4$ and $I = 1$ empirically.

7.6 Preparing Data for Empirical Induction

Teacher candidates were then asked to conjecture what the areas might be for other cells and to test their conjectures on a computational geoboard. Their conjectures were based on recognizing the recursive dependence between two neighboring cells and described in words in two different ways. Namely, some of the teacher candidates came up with the wording like "add one-half as you go down" thus recognizing the vertical pattern, whereas others came up with "add one as you go across" thus recognizing the horizontal pattern. At this point, one teacher candidate raised the question: "Which rule is correct?", wondering whether different rules may yield the same numbers and if so, how a spreadsheet can be used to justify the equivalence of the two different rules (wordings) through numerical evidence.

In response, the class was shown that these intuitive rules can be effectively communicated to the spreadsheet through action (cell referencing). Such action involves (a) typing the equal sign in cell C3, (b) clicking cell B3 (or C2), (c) typing the plus sign in cell C3, (d) typing 0.5 (or 1) in cell C3, and (e) pressing the *enter* key. The use of pointing (with simultaneous mouse clicking) allowed the teacher candidates to shift the burden of *symbolization* onto the spreadsheet. The instructor emphasized that the software is able to do more, however, than to ascribe meaning to concrete action, in other words, more than to translate the pointing into a symbolic formulation. The software is able to replicate the concrete action with respect to a specific pair of cells to any such pair. In doing so, one can shift also the burden of *generalization* onto the software by replicating the content of cell C3 to cell H8. Fig. 7.11 and Fig. 7.12 show the results of such replication based on 'vertical' and 'horizontal' rules, respectively, with their symbolic representations (spreadsheet formulas) displayed in the *Formula Bar*. Therefore, spreadsheet modeling made it possible to justify the equivalence of the two rules through numerical evidence.

In such a way, the computational experiment aimed at preparing data needed for inductive discovery of the area-on-geoboard concept also allowed teacher candidates to use empirical results as a convincing justification of their emerging mathematical curiosity. The main role of the spreadsheet, however, is to mediate inductive reasoning; in other words, to aid the teacher candidates in abstracting from the specificity of each individual number to the generality of a relationship among the numbers. Such a relationship can be first expressed in a pictorial form, something that provides a visual support for understanding algebraic symbolism.

C3				fx	=C2+0.5		
A	B	C	D	E	F	G	H
B\I	**0**	**1**	**2**	**3**	**4**	**5**	**6**
3	0.5	1.5	2.5	3.5	4.5	5.5	6.5
4	1	2	3	4	5	6	7
5	1.5	2.5	3.5	4.5	5.5	6.5	7.5
6	2	3	4	5	6	7	8
7	2.5	3.5	4.5	5.5	6.5	7.5	8.5
8	3	4	5	6	7	8	9
9	3.5	4.5	5.5	6.5	7.5	8.5	9.5

Fig. 7.11. Template filled through 'vertical' rule: C3 = C2 + 0.5.

C3				fx	=B3+1		
A	B	C	D	E	F	G	H
B\I	**0**	**1**	**2**	**3**	**4**	**5**	**6**
3	0.5	1.5	2.5	3.5	4.5	5.5	6.5
4	1	2	3	4	5	6	7
5	1.5	2.5	3.5	4.5	5.5	6.5	7.5
6	2	3	4	5	6	7	8
7	2.5	3.5	4.5	5.5	6.5	7.5	8.5
8	3	4	5	6	7	8	9
9	3.5	4.5	5.5	6.5	7.5	8.5	9.5

Fig. 7.12. Template filled through 'horizontal' rule: C3 = B3 + 1.

7.7 Abstracting from Numbers to Equations Using First-Order Symbols

With this in mind, teacher candidates were introduced to the use of the first-order symbols in the form of differently shaded boxes (cells) as iconic representations of numerical patterns generated by a spreadsheet. The wordings "add one as you go across" and "add one-half as you go down" can be represented in a diagrammatic form as support system in appreciating variables as the major elements of algebraic formalism. Fig. 7.13 and Fig. 7.14 show iconic representations of 'vertical' and 'horizontal' rules, respectively. The first step towards abstracting from numerical patterns to algebraic equations is to use the first-order symbols for re-formulating the two wordings as follows: 'dark cell equals light cell plus one-half' (Fig. 7.13) and 'dark cell equals light cell plus one' (Fig. 7.14).

Shaded cells like variables do not depend on specific numbers and express relations of generality at the level of the first-order symbols. This suggests that the use of such symbols mediates the transition from empirical evidence of the recursive variation of areas to a symbolic form of the recursion. One can denote $A(B, I)$ the area of a polygon with B border and I interior points and use this notation in writing down the 'vertical' and the 'horizontal' definitions for area-on-geoboard, respectively:

$$A(B, I) = A(B-1, I) + 0.5 \tag{7.2}$$

and

$$A(B, I) = A(B, I-1) + 1, \tag{7.3}$$

provided $A(3, 0) = 0.5$.

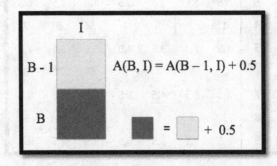

Fig. 7.13. Visualization of the 'vertical' rule.

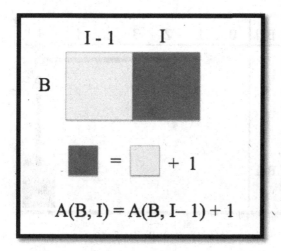

Fig. 7.14.   Visualization of the 'horizontal' rule.

7.8   Visual and Symbolic Deduction of Pick's Formula

The structure of a spreadsheet allows for the use of visualization in guiding inductive discovery of discrete concepts through manipulating cell-patterned diagrams. Indeed, as the diagrams in Fig. 7.13 and Fig. 7.14 show, one can visualize area-on-geoboard concept; that is, one can read out the meaning of Eq. (7.2) and Eq. (7.3) from the diagrams of Fig. 7.13 and Fig. 7.14, respectively. Then, one can use the visualization as a means to deduct a closed formula for the area $A(B, I)$ from its recursive definition.

For teacher candidates with limited mathematical experience in formal mathematics, the distinction between recursive and closed formulas, or better the *meanings* of the formulas, is wrapped up tightly into the cumbersome terminology. As a way around this difficulty, one can use mathematical visualization to carry the meanings from one formula to another. More specifically, one can use numerical evidence to demonstrate two different ways of generating the content of any cell of the spreadsheet: (i) through one of the neighboring cells (Fig. 7.15 and Fig. 7.16), and (ii) through the far top cell (I-row) and the far left cell (B-column) of the spreadsheet, that is, through I and B (Fig. 7.17).

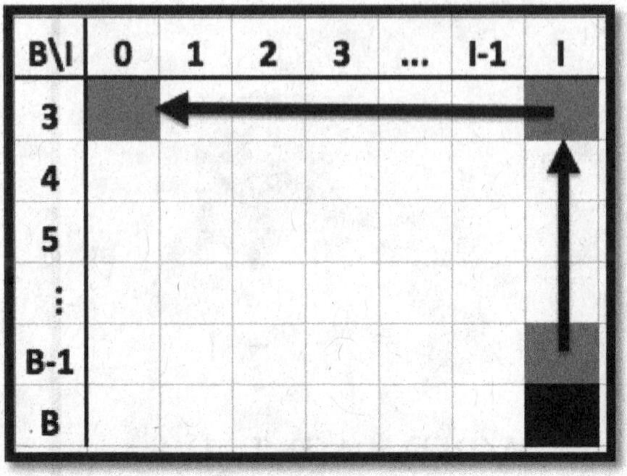

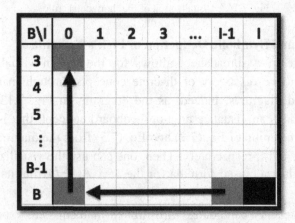

Fig. 7.15. Calculating $A(B, I) - A(3, 0)$: $B - 3$ steps up and $I - 0$ steps to the left.

Fig. 7.16. Calculating $A(B, I) - A(3, 0)$: $I - 0$ steps to the left and $B - 3$ steps up.

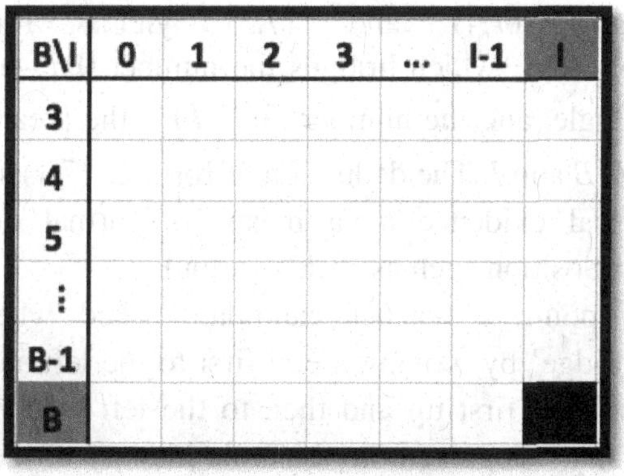

Fig. 7.17. Visualization of a closed-form formula.

The demonstration began with the use of the diagram of Fig. 7.13 according to which, starting from any two vertically adjacent cells, one can replace the upper cell by the cell immediately above, plus one-half. In other words, the first term of the right-hand side of Eq. (7.2), namely, $A(B-1, I)$, can itself be replaced by $A(B-2, I)+0.5$. Fig. 7.15 shows how the movement of the grey cell through its subsequent replacement by the cell immediately above links the dark cell to the top cell on the chart's boundary and step by step accumulates one-halves while numbers in the B-column mediate the counting of the steps. Once the top cell has been reached, the diagram of Fig. 7.14 comes into play allowing for a horizontal movement of the grey cell to the left. Each step to the left accumulates unity while numbers in the I-row count the steps. Thus, one can visually derive closed formula from a recursive definition (formula) of area by linking an arbitrary cell (area) to the top left cell (area of the fundamental triangle) of the spreadsheet. This visual description of the transformations of cells can be represented in a symbolic form as follows.

Vertical movement:

$$A(B, I) = A(B-1, I) + 0.5 = A(B-2, I) + 2 \cdot 0.5$$
$$= \ldots = A(3, I) + (B-3) \cdot 0.5 .$$

(7.4)

Horizontal movement:

$$A(3, I) = A(3, I-1) + 1 = A(3, I-2) + 2 \cdot 1$$
$$= \ldots = A(3, 0) + I \cdot 1 = 0.5 + I .$$

(7.5)

Combining (7.4) and (7.5) yields formula (7.1), $A(B, I) = 0.5 \cdot B + I - 1$, which bridges the number 0.5 – the area of the fundamental triangle, and the number $A(B, I)$ – the area of an arbitrary polygon, through B and I. The deduction of formula (7.1) shows how one can use numerical evidence as a means of formal deduction of a mathematical proposition such as Pick's formula.

At this point, a teacher candidate asked whether one can construct this 'bridge' by moving a cell first to the left and then up (see Fig. 7.20) rather than first up and then to the left (see Fig. 7.19). She could have visually recognized the dynamic identity of the 'up–left' and 'left–up' movements, and the inquiry was, probably, an attempt to share this discovery with others. This eureka-like inquiry, however, might have been caused by the lack of her mathematical experience with a situation in which the same result can be developed in more than one way; for example, through either horizontal or vertical movement, as above. However, the use of the spreadsheet's structure as a didactic tool in visualizing different ways of bridging the two cells, has transformed the teacher candidate's mental functioning in a way that allowed for the discrimination of a mathematical idea that permeates a formal deduction of the closed formula for the area-on-geoboard concept. Yet, such discrimination occurred before she really understood how this idea can be expressed in an algebraic form. Here, one can recognize the emergence of mathematical meaning through visualization: as a result, the teacher candidates were able to demonstrate their awareness of the commutativity of the up and left movements in subsequent independent explorations (see below Fig. 7.20 and Fig. 7.21).

To conclude this section, note that the use of a spreadsheet as a tool of visualization mediated one's transition from the recursive formulation of area to its closed form in which visualization provided a support system in articulating formal deduction. Furthermore, the use of a spreadsheet as a computational tool allowed for connecting Pick's formula with its empirical justification; that is, by computing areas through formula (7.1) the teacher candidates were able to arrive at the same array of numbers (areas) that was used by them as a mediator of deductive reasoning. Thus, the duality of spreadsheet's functioning enables one to utilize both inductive and deductive reasoning in the context of informal geometry.

7.9 Moving to a New Learning Site

A rich didactic interplay that was found between Pick's formula and the use of technology makes it possible to introduce another mathematically significant topic which provides teacher candidates with a similar structure of explorations. The similarity of explorations allows for connecting old and new contents also. As will be shown below, the same (technology-enabled) method would not only open a window to a different content but better still, demonstrate the equivalence of the contents. Through extended engagement in computer explorations one can apply skills and strategies acquired in the context of re-discovery Pick's formula to a new learning situation dealing with a famous formula by Euler

$$F + V - E = 1, \tag{7.6}$$

connecting the number of faces (F), vertices (V), and edges (E) in any polygon. On a geoboard, formulas (7.1) and (7.6) are uniquely related: they turn out to be equivalent [DeTemple and Robertson, 1974].

On a geoboard, any polygon can be triangulated by connecting border and interior points which then become the nodes of the so triangulated figure. The concept of vertices can then be extended to these nodes. Like the area, three other characteristics – F, V, and E – of a triangulated polygon can be introduced and can be expressed in terms of B and I. Using computational geoboards, one can explore numerically the dependence of E on B and I in much the same way as in the case of area. (Unfortunately, *The Geometer's Sketchpad* does not provide the reliable count of selected segments on a constructive basis; yet the counting can be done directly by a user.) Furthermore, F and V allow for a simple representation in terms of B and I, provided that Pick's formula is known. The combination of F, V, and E expressed in terms of B and I yields Euler's formula for a polygon on a geoboard.

The third part of the activities dealt with the teacher candidates' independent investigation aimed at the re-discovery of Euler's formula through Pick's formula and establishing the equivalence of the two propositions on a geoboard.

Once again, the author's use of technology made it possible to introduce the subject of explorations to the teacher candidates. This included the discussion of the idea of triangulation of a lattice polygon and how one can use *The Geometer's Sketchpad* in triangulation, the introduction of the triple (V, F, E) as fundamental characteristics of a triangulated polygon. Then the teacher candidates were given an assignment which included several tasks. The following four sections present the tasks, responses to them by different teacher candidates and the authors' analysis of their results.

7.10 Measuring vs. Counting

In order to explore the dependence of V and F on B and I the following two tasks were designed.

Task 4. *Construct a polygon on a computational geoboard. Count the number of border and interior points; that is, find B and I for your polygon. Then triangulate the polygon. Count the number of vertices of the triangulated polygon; that is, find V. How are V, B, and I related? Write down this relation. Test your findings for several other polygons.*

Task 5. *Using your triangulated polygon count the number of faces (fundamental triangles); that is, find F. Recall that the area of the fundamental triangle is one-half. How are F, B, and I in your polygon related? Write down this relation.*

One may note that Task 5 is designed in a way which does not encourage measuring; one is even reminded what area of the fundamental triangle is. Yet, as the result shown in Fig. 7.18 indicates, a teacher candidate did not take advantage of the reminder. Rather, she preferred to use measuring as a method of developing an empirical evidence to support inductive reasoning. It is this reasoning that allowed her to spot a relation between the area of a polygon and the number of faces resulted from its triangulation. The idea that the number of faces is twice the area can be grasped on a single polygon; however, the grasp of this idea outside a computational environment involves at least three different components of mathematical thinking: geometric intuition, coordination of the dimensional character of *area* and the dimensionless of *number* (the number of faces), and the recognition of the additivity of area. So, despite the wording of Task 5 which focuses on a single polygon, the teacher candidate worked through a sample of triangles and, like in the case of Pick's formula, chose to take advantage of the computational geoboard. The use of the results of measurement as support system in making the induction was of help in thinking about how to link geometric and symbolic domains. Once again the *mechanical* method of finding areas first helped in grasping a relation between measured and counted objects within a numerical domain and then assisted in moving to the symbolic domain in formulating a new geometric result in terms of a familiar concept.

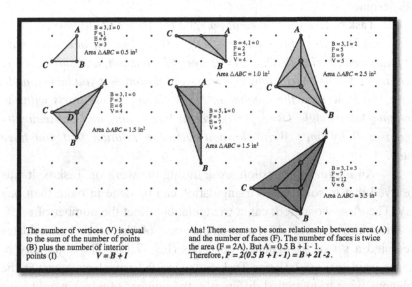

Fig. 7.18. Representative sample for spotting patterns.

The result shown in Fig. 7.18 demonstrates the use of both inductive and deductive reasoning within a medium of exploration and the articulation of a formal demonstration of the existence of a non-trivial geometric relationship. The use of 'Aha!' can be perceived as an unconscious distinction between the levels of complexity of the results of Task 4 and Task 5. While the former task involved the adding of parts to get a whole and thus was ascertained by the teacher candidate with a relative ease, the latter task required cognitive efforts that went far beyond the immediate recognition of a whole-part relationship.

Another observation common to all the activities is that although the tasks suggest constructing *any* lattice polygon, most of the teacher candidates once again stuck with triangles in their explorations like the one whose result is displayed in Fig. 7.18. However, it is the simplicity of triangles that allowed her to vary their sizes in order to provide a representative sample for constructing empirical evidence needed for successful inductive discovery.

7.11 Encountering Limitation of the Environment

In order to explore the dependence of E on B and I the following task was designed.

Task 6. *Use your original polygon to explore the following:*

a) How does the number of edges (E) change if one starts with the fundamental triangle (B=3, I=0), keeps B constant, and subsequently increases I by one? Write down a recursive relation that you have found.

b) How does the number of edges change if one starts with the fundamental triangle (B=3, I=0), keeps I constant, and subsequently increases B by one? Write down a recursive relation that you have found.

An interesting question arose during the work on Task 6. It was observed that in some cases triangulation can be done in more than one way. The class wondered: can a triangulation affect the number of edges in a resulting figure? As an example, one of the teacher candidates presented a set of triangulated polygons (Fig. 7.19). The simplicity of shapes used in Fig. 7.19 for Task 6b suggested trivial evidence: if the polygons were triangulated differently, the number of edges would have stayed the same. It has also appeared that triangulations of the triangles in Task 6a are the only possible ones. Extending visually the far right triangle to a few more interior points seemed not to bring about a counter-example. But the ease of justification of emerging conjectures experienced through measuring of areas in the context of Pick's formula called for more empirical evidence. However, as it was already mentioned, *The Geometer's Sketchpad* does not allow for an automatic counting of selected objects on a constructive basis.

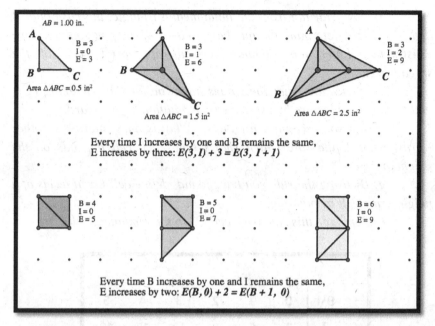

Fig. 7.19. Spotting numerical patterns through limited empirical evidence.

Thus, in an attempt to gather more empirical evidence the class has unexpectedly reached the limitations of the environment: the by-hand counting of a large number of objects did not seem to be an effective way of investigating the situation in a computational media. Exploring the new questions (with obvious mathematical depth) turned out to be a challenge. It was decided that one can proceed with explorations further by building a conclusion about the invariance of the number of edges under a triangulation on limited empirical evidence. So, Fig. 7.19 shows how the teacher candidate constructed the boundary conditions for edges through such evidence. Therefore, one may distinguish between generating boundary conditions for areas through measurement and generating boundary conditions for edges through counting in terms of the level of empirical substantiation.

7.12 From Particular to General through Visualization
Task 6 continues.

    c) Enter your data into the spreadsheet provided.

d) Now, starting from the fundamental triangle, increase both B and I by 1 and determine the number of edges of the polygon you have constructed. Enter this number into an appropriate cell of the spreadsheet.

e) Conjecture what the values might be for the remaining cells and test your conjecture by using the (computational) geoboard.

f) Can you recognize a pattern? What is this pattern? In other words, what is the rule which shows how each cell depends on the neighboring cell?

g) By using the rule you have found, define cell C3 in terms of a neighboring cell by pointing.

h) Replicate this rule from cell C3 to the remaining cells.

C3				fx	=C2+2	
	A	B	C	D	E	F	G
1	B\I	0	1	2	3	4	5
2	3	3	6	9	12	15	18
3	4	5	8	11	14	17	20
4	5	7	10	13	16	19	22
5	6	9	12	15	18	21	24
6	7	11	14	17	20	23	26
7	8	13	16	19	22	25	28
8	9	15	18	21	24	27	30
9	10	17	20	23	26	29	32

Fig. 7.20. Edges through 'vertical' recursion: C3 = C2 + 2.

C3		⇕	⊗	✓	⟲	*fx*	=B3+3

	A	B	C	D	E	F	G
1	**B\I**	**0**	**1**	**2**	**3**	**4**	**5**
2	**3**	3	6	9	12	15	18
3	**4**	5	8	11	14	17	20
4	**5**	7	10	13	16	19	22
5	**6**	9	12	15	18	21	24
6	**7**	11	14	17	20	23	26
7	**8**	13	16	19	22	25	28
8	**9**	15	18	21	24	27	30
9	**10**	17	20	23	26	29	32

Fig. 7.21. Edges through 'horizontal' recursion: C3 = B3 + 3.

Fig. 7.20 and Fig. 7.21 represent two spreadsheets developed through independently completing Tasks 6c through 6h by a teacher candidate who, in generating edges, used 'vertical' and 'horizontal' recursions, respectively.

i) Write a general (closed) formula for the number of edges E of a triangulated polygon with B border points and I interior points.

Fig. 7.18 shows how visualization provided by the spreadsheet of Fig. 7.20 and thinking in terms of steps needed to connect an arbitrary chosen cell with the top left cell, mediated the transition from numeric to symbolic domains in exploring the last question. Although the deduction of the Pick-like formula for edges has not been accompanied by a mature use of algebraic symbolism, the report by the teacher candidate presented below clearly indicates her appreciation of a spreadsheet as a mediator of formal deduction of a mathematical proposition at the appropriate level of rigor. The use of the word *any* with an immediate reference to the specific choice of a cell, illustrates how the environment brings about the existence of an equilibrium dynamic between general and particular in the process of the teacher candidate's deductive reasoning. Using the spreadsheet pictured in Fig. 7.20, the teacher candidate reported as follows.

i) The formula for finding the number of edges in term [sic] of interior points and border points is:

First, choose any cell in your template. I chose 25 with B = 8 and I = 4.

$25 = 15 + 2\,(\text{\# of steps}).$

$\bigstar\, 25 = 15 + 2 \cdot (B - 3).$

Then

$15 = 3 + 3(\text{\# of steps}).\ \ 15 = 3 + 3 \cdot I$

Substituting $3 + 3 \cdot I$ *into* $\bigstar$ *for 15 we get*

$25 = 3 + 3 \cdot I + 2 \cdot (B - 3) = 3 \cdot I + 2 \cdot B - 3.$

Therefore,

$E = 3 \cdot I + 2 \cdot B - 3.$

7.13 Communicating about Mathematics

Task 6 continues.

j) After you have formulated V, F, and E in terms of B and I, write down a relation connecting V, F, and E. You are supposed to obtain a very remarkable relation known as Euler's formula for a lattice polygon. Discuss why the relation which you have discovered is remarkable.

k) Derive Pick's formula from Euler's formula. Discuss the notion of equivalence of the two statements.

The teacher candidate continued her report by answering the two questions as follows.

j) What is the relationship between F, V, and E?

We have:

$$V = B + I,\ F = 2A = 2(0.5B + I - 1) = B + 2I - 2,\ E = 3I + 2B - 3.$$

From here it follows:

$$I = V - B,\ F = B + 2(V - B) - 2 = 2V - B - 2,\ B = 2V - F - 2\,.$$

Also

$$B = V - I,\ F = V - I + 2I - 2,\ I = F - V + 2\,.$$

Finally,

$$E = 3(F - V + 2) + 2(2V - F - 2) - 3$$
$$= 3F - 3V + 4V - 2F + 6 - 4 - 3 = F + V - 1\,.$$

Therefore, $F + V - E = 1$ *– Euler's Theorem!!!*

This relationship between F, V, and E is "remarkable" because it is constant. The constant in this relationship is 1. No matter what size or shape of polygons constructed, the relationship between these three variables will always be the same. This would be a "special" characteristic for all polygons.

k) $F + V - E = 1, 2A + (B + I) - (3I + 2B - 3) = 1, 2A = -2 + B + 2I$.

Therefore, $A = 0.5B + I - 1$ *– Pick's Theorem.*

If two theorems are equivalent, as in the case of Pick's and Euler's Theorems (arrived by different means and times exclusive from one another), then it comes to be seen that this type of "connection" between two or more concepts is a very rare human accomplishment. It seems as though most concepts which can be derived from one another (i.e., area of parallelogram ... to area of rectangle) form a sort of hierarchy from generalized to more specific ideas (i.e., all rectangles are parallelograms but not all parallelograms are rectangles). A particular vertical "order" or "classification" of concepts seems to be the dominant structure for most mathematical relationships. Pick's and Euler's Theorems defy this structure, as they exist "separate but equivalent", thus forming a more horizontal structure within their relationships.

The teacher candidate's communication about the nature of mathematics indicates her confirmation of a kind of ownership over two big mathematical ideas due to an active participation in the process of re-discovery enhanced by the use of technology. Having experience with doing mathematics, such sense of ownership over famous mathematical formulas turned out to be thought provoking for the teacher candidate and it served as an inspiration to discuss the structure of mathematics in rather general terms. As an extension of this discussion, one can also be reminded that measuring lies at the origin of mathematics and, therefore, it has to be considered a significant activity in the technology-supported study of geometry by teacher candidates. Measuring is a simple method of empirical analysis of geometric objects and it brings about empirical evidence with a relative ease. This is the main reason why the method is so attractive for the users of dynamic geometry environments. However, the role of empirical evidence in the development of geometry was not to set up properties of geometric objects, but rather to generalize first experience to concepts and then make use of the concepts for formal demonstration of their properties. That is, in the spirit of Archimedes, first to make sense of concepts by a "mechanical method" and then demonstrate their properties by formal means. The necessity of multiple measurements for developing the area-on-geoboard concept from empirical evidence made it possible to shift the emphasis of explorations from measuring alone to a variety of computing activities using different digital tools.

Due to the equivalence of Euler's and Pick's formulas on a geoboard, the two famous propositions have a specific appeal for mathematics education. Their statements can be understood at the elementary level, yet the mathematical content offers collateral learning opportunities opening a window to such important concepts as recursion, triangulation, invariance, and equivalence, to name a few. The appropriate use of technology can help one understand these concepts by intertwining action, measurement, visualization, conjecturing, inductive and deductive reasoning within a computational medium. Finally, acquiring mathematical knowledge through the process of re-discovery can support learning mathematical ideas in a natural order following the general principle that "the best way to guide the mental development of the individual is to let him retrace the mental development of the race" [Ahlfors, 1962, p. 190]. The use of computing technology in mathematics teacher education significantly facilitates the process of re-discovery mathematical results and, in some cases, can lead to the discovery of new mathematical knowledge [Abramovich and Leonov, 2009].

Chapter 8

Probability as a Blend of Theory and Experiment

8.1 Introduction

Probability is a curricular strand the teaching of which begins in the early grades and continues through secondary and tertiary mathematics education. When preparing to teach the concepts of probability, elementary teacher candidates can further develop expertise they "should seek to develop in their students" [Common Core States Standards, 2010, p. 6] as they learn the standards for mathematical practice associated with this part of the curriculum. These standards emphasize mathematical modeling, the appropriate use of physical and digital teaching tools, and the need to pay attention to precision in problem solving. There are many problem situations motivated by the standards and associated with uncertain (random) outcomes. Put another way, such problems emphasize the "ideas of uncertainty to illustrate that mathematics involves more than exactness when dealing with everyday situations" [New York State Education Department, 1998, p. 48].

For example, a child tosses a coin and knows with certainty that it would not fly but fall as a result (the child might request explanation of this seemingly simple phenomenon); yet he/she cannot predict with certainty how exactly it would fall: head or tail. A more senior experimentalist can somehow attempt to measure each of the two possible outcomes. Note that measurement is one of the basic human endeavors that gave birth to numbers. In turn, it is perhaps early experiments with random outcomes such as an ancient game of guessing the parity (odd or even) of objects hidden from view [Smith, 1953] that had been responsible for assigning names to numbers. Much later, in the 18th century, mathematical perspectives on this popular game have been a source of disagreement among notable creators of the theory of probability (see below section 8.7).

A classic definition of the concept of probability, commonly accepted nowadays in school mathematics, was introduced by Cardano[18] as the ratio of the number of favorable incidents to the total number of equally likely incidents within a certain observation [Tijms, 2012, p. 2]. This definition shows that solving problems with equally likely random outcomes gives an applied flavor to proper fractions as numerical characteristics of what is considered probable and makes it possible to compare chances (the measures of which are called probabilities) by using the second-order symbols (i.e., rational numbers in the range [0, 1]). In that way, one can learn to "make sense of quantities and their relationships in problem situations" [Common Core States Standards, 2010, p. 6] that do not have certain outcomes. In the words of Descartes[19] [1965], "When it is not in our power to determine what is true, we ought to act according to what is most probable" (p. 21). To understand what is most probable; in other words, to make sense of measuring the likelihood of uncertain (random) outcomes, one must be able to compare the probabilities of such outcomes. But probabilities (or chances), in order to be compared, have to be computed (measured) first. Mathematical actions of that kind should not be considered through the simplistic lenses, like measuring perimeters and/or computing areas of basic geometric shapes.

The complexity of computing probabilities (or measuring chances) deals in part with the fact that there are many real-life situations when random outcomes are not equally likely and therefore, probabilities (chances) can only be determined experimentally. Here is a simple example. Only in an idealized situation that we casually say there are two chances out of five to pick up (without looking) an apple from a basket with two apples and three pears. But in a real-life, a basket may have tiny apples and large pears (or vice versa or all the fruits having distinctly different size). In that case, the chances to pick up an apple can only be determined experimentally by reaching into the basket without looking, say, 100 times and record the frequency of apples to be drawn through this experiment. (Note that here we consider 100 as a large number of trials). Furthermore, even knowing that out of 100 trials, we ended up

---

[18] Gerolamo Cardano (1501–1576) – an Italian mathematician.

[19] René Descartes (1596–1650) – a French mathematician and philosopher.

with an apple 10 times, the most a beginning learner of the probability strand can say is that we have observed 10 outcomes out of 100 total in favor of an apple. This informal conclusion about the number of successful outcomes does not offer a numerical value in order to determine whether in the case of another basket (or, more generally, situation) the success is more (or less) probable. In some simple cases, like having one basket with a few apples and another one with a lot of apples, this comparison may be carried out intuitively at the level of the first-order symbols. But not all cases are that simple. Therefore, a motivation for learning to compute probabilities at the level of the second-order symbols may be due to the need to compare them, so that in a real life to be able "to act according to what is most probable" [Descartes, 1965, p. 21]. But as always, the use of the first-order symbols should precede the use of the second-order symbolism. This brings about an idea that chances can be expressed through the process of geometrization of an experimental situation with random outcomes and a large number of trials (experiments) from where measures of the corresponding geometric images can be derived. That is how young children have been taught to compare whole numbers using their geometric representations in the form of towers built out of identical square tiles. Through this comparison, the idea of the unit of measurement, a tile, has been considered critical.

This chapter will revisit from a probabilistic perspective a number of problems from the previous chapters and consider some new (for this book) problems. A strong emphasis will be given to experimentally determining probabilities (or, better, relative frequencies) as the ratio of the number of trials a desired event occurs to the total number of trials within an experiment. To this end, several means of experimentation will be suggested and discussed. These will include a combination of computing technology and hands-on paper-and-pencil explorations.

8.2 Basic Concepts and Tools of the Probability Strand

8.2.1 *Randomness and sample space*

In this section, several basic concepts associated with the probability strand will be explained. The first one is randomness – a characterization of the result of an experiment that is not possible to predict but, nonetheless, is often possible to measure on the scale from zero to one, with zero assigned to something impossible (e.g., a coin flies when tossed in the still air) and one assigned to something certain (e.g., a coin falls when tossed in the still air). The very concept of randomness is not easy to define; it "has proved to be more than a little elusive, and it remains enigmatic, despite the fact that it has proved to be useful in many contexts" [Nickerson, 2004, p. 54]. One can say that randomness does not lead to a credible pattern, although it is often difficult to conclude whether there is no pattern in a sequence of events. Nonetheless, thinking about randomness in a very informal way, can help one understand the "difference [between] predicting individual events and predicting patterns of events" [Conference Board of the Mathematical Sciences, 2001, p. 23]. When tossing a coin, one cannot predict how exactly it would fall let alone to predict a pattern in the sequence of tosses, although it is not impossible to have head and tail alternating in, say, five tosses, or even to have five heads or tails in a row. Whatever a prediction, the question to be answered is how to measure the likelihood (chances) of the prediction.

The need for measuring chances leads to the concept of the sample space, which is considered difficult for young children as, frequently, the outcome of data collection depends on a population involved or a context considered [National Council of Teachers of Mathematics, 2000]. The sample space of an experiment with a random outcome is the set of all possible outcomes of this experiment. Within a sophisticated experiment, for young children (as well as for their future teachers) it is difficult to determine whether such a set is complete because the confirmation of completeness requires solving a separate mathematical problem. Such problems may differ significantly in terms of complexity. A simple example is the sample space of rolling a six-sided die (with the number of spots on the sides ranging from one to six) comprised of six outcomes: {1, 2, 3, 4, 5, 6}. Assuming that we deal with a fair (unbiased) die, all outcomes may be considered equally likely. Under this assumption, one can say that there is one chance out of six to cast any of the six numbers (spots).

A more complicated example is the sample space of an experiment of changing a quarter into pennies, nickels, and dimes. The chart of Fig. 8.1 shows a 12-element sample space, where numbers in the top row and the far-left column show the ranges for dimes and nickels, respectively, and the remaining numbers show the corresponding number of pennies in a change. For example, the triple (5, 2, 1) stands for five pennies, two nickels, and one dime, so that $5 \cdot 1 + 2 \cdot 5 + 1 \cdot 10 = 25$. Alternatively, the sample space can be described as the following set of triples of numbers {(25, 0, 0), (20, 1, 0), (15, 2, 0), (10, 3, 0), (5, 4, 0), (0, 5, 0), (15, 0, 1), (10, 1, 1), (5, 2, 1), (0, 3, 1), (5, 0, 2), (0, 1, 2)}, where each triple describes the number of pennies, nickels, and dimes, respectively, in a change. Yet another, quite cumbersome, representation of the sample space will be shown below. Note that there is no reason to assume that it is equally likely to get any combination of the coins from a change-making device and Fig. 8.1 represents a sample space with not equally likely outcomes. Therefore, similarly to the case of a basket with fruits, given a change-making device, it is only experimentally that one can determine chances for a specific combination of coins in a change.

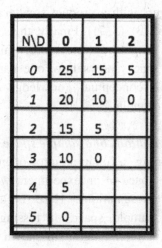

N\D	0	1	2
0	25	15	5
1	20	10	0
2	15	5	
3	10	0	
4	5		
5	0		

Fig. 8.1. Organizing data in a chart.

8.2.2 *A more complicated example of constructing a sample space*

A somewhat more complicated example deals with selecting natural numbers not greater than, say, 40 with exactly three different representations as a sum of consecutive natural numbers (Chapter 6, section 6.9). Note that the first example requires basic conceptual understanding of what is possible through rolling a die. The second example requires understanding of a more complicated concept dealing with a system of partitioning an integer, 25, into three specified summands, 1, 5, and 10. The example of this section requires advanced conceptual understanding of how to find the number of trapezoidal representations of a natural number. To this end, one can use Remark 6.3, Chapter 6, stating that the representation of a natural number in the form $mn/2$, where $m > n$ are integers of different parity, implies that this number has a trapezoidal representation with n rows. Let $m = 2k$, then $mn = 2kn$. Note that when k and n are prime numbers, the product $2kn$ can be represented as a product of two factors in three ways only: $(2k)n$, $k(2n)$, and $2(kn)$. In other words, three prime numbers including the number 2 can form only three different pairs of factors. However, if one of the numbers n or k is a product of two primes, then two factors can be formed in more than three ways. Indeed, let $n = pq$ where $p > q > 1$ are prime numbers. One can check to see that there exist seven ways to arrange the product $2kpq$ as the product of *two* factors. Therefore, the following pairs of integers with a product smaller than 40 are possible: $\{(3, 5), (3, 7), (3, 11), (3, 13), (5, 7)\}$. Finally, the set $\{15, 21, 33, 35, 39\}$ is the sample space sought. In particular, this third example of constructing a sample space shows how one can connect procedural knowledge (formula) to conceptual knowledge necessary to extract hidden meaning from a formula.

8.2.3 *Different representations of a sample space*

As the case of changing a quarter into other coins has already demonstrated, a sample space of an experiment with random outcomes may have different representations. Here is another, conceptually distinct, example: the sample space of tossing two coins can be represented in the form of a tree diagram as shown in Fig. 8.2. Note, that the outcomes of the two tosses are independent, that is, the outcome of the second toss does not depend on what happened on the first toss. This independence is reflected in the very form of the tree diagram: each of the two possible outcomes of the first toss affords the same two outcomes for the second toss. This, however, is not always the case and in more complex situations, drawing a tree diagram to represent a sample space may be a difficult proposition.

An outcome of an experiment is an element of the corresponding sample space. For example, the sample space of the experiment of tossing two coins shown in Fig. 8.2 consists of four outcomes. Further, outcomes may be combined to form an event. For example, the outcomes Head-Head and Tail-Tail form an event that both coins turn up the same. Outcomes may or may not be equally likely, something that, of course, depends on the conditions of an experiment. All the outcomes with tossing coins or rolling dice described in the above examples are considered equally likely outcomes assuming that we deal with the so-called fair coins of fair dice. An assumption about equally likely outcomes is a theoretical assumption – after all, when we deal with a coin (or die), we *assume* that it is a fair coin (or die).

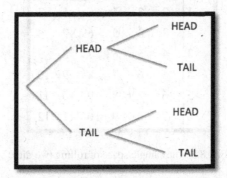

Fig. 8.2. The sample space of tossing two coins.

Outcomes are not always equally likely. Indeed, when having a bag with three red M&Ms and two yellow M&Ms, it is not equally likely to pick up either a red or a yellow candy. Likewise, events may be independent and dependent. In the case when from the first draw of an M&M from the bag, the candy is returned to the bag, the event of picking up a red M&M on the second draw does not depend on which color candy was picked up on the first draw. At the same time, if the candy is not returned, the event of picking up a red M&M on the second draw depends on what happened on the first draw.

The sample space of rolling two dice and recording the total number of spots on two faces can be represented in the form of an addition table (Fig. 8.3) from where it follows that the highest chances are for having seven as the total number of spots on two dice. Here one can connect probability concepts to the concepts of integer partitions. Indeed, whereas ten has more partitions than seven, in the context of rolling two dice partitions are limited to the summands that are not greater than six. Under such condition, the closer a number to six is, the more ordered partitions into two integer summands exist. For example, the sum 12 has only one possibility and the sum 7 has six possibilities (Fig. 8.3). This shows the importance of paying attention to linguistic coherency (Chapter 3, section 3.2.3) in the process of formulating/posing mathematical tasks.

	1	2	3	4	5	6
1	2	3	4	5	6	7
2	3	4	5	6	7	8
3	4	5	6	7	8	9
4	5	6	7	8	9	10
5	6	7	8	9	10	11
6	7	8	9	10	11	12

Fig. 8.3. The sample space of rolling two dice.

Finally, consider a tree diagram representation of the sample space of the experiment of changing a quarter into three smaller coins. In fact, teacher candidates often ask a question whether one can carry out the finding of the number of ways to change a quarter by using a tree diagram. Note that unlike the cases of tossing coins or rolling dice, in the case of getting change one deals with outcomes in the form of the triplets of integers being in relation of mutual dependency. Indeed, the number of dimes in a change depends on the number of nickels and both depend on the number of pennies. Whereas it should not be recommended to use a tree diagram for constructing the sample space of this problem, notwithstanding, it can be constructed using this method. As shown in Fig. 8.4, the form of the tree diagram reflects the notion of interdependency of possible outcomes and the didactic value of representing a sample space through a tree diagram is in demonstrating the logic of causality involved in the construction of the tree.

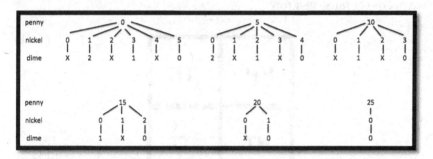

Fig. 8.4. The diagram reflects the dependence of possible outcomes in making a change.

8.3 Fractions as Tools in Measuring Chances

The application of arithmetic to geometry (discussed in the context of the triangle inequality in Chapter 3) brings about the idea of using a numeric measure of the likelihood of an event. As was mentioned in Chapter 4 (section 4.5.1), one of the ways to introduce the concept of fraction as an applied tool is through using fractions as a measure of the likelihood (or probability) of an event. The idea of a geometric representation of a fraction enables its use as a means of finding the probability of an event (alternatively, measuring its likelihood).

Consider equally likely outcomes of the experiments with tossing a coin and rolling a die. The meaning of the words *equally likely* can be given a geometric interpretation through representing such outcomes as equal parts of the same whole. In the case of coins, we have four equally likely outcomes, which can be represented in the form of a rectangle divided into four equal parts so that each part is represented by a unit fraction 1/4. It is this fraction that can be considered as the value of the likelihood (probability) of any of the four (equally likely) outcomes. Dividing the rectangle in four equal parts as shown in Fig. 8.5

demonstrates two things: (i) a table representation of the sample space of tossing two coins; (ii) according to the area model for fractions (Chapter 4), each of the four parts represents the product $(1/2)(1/2)$, where each factor is the probability of head/tail for each of the two tosses. Extending the number of tosses beyond two, one can measure the probability of alternating heads and tails in, say, five tosses by the number $(1/2)^5$. One can see that the same number also measures the probability of having five heads (or tails) in a row.

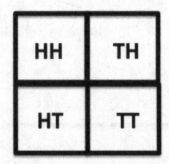

Fig. 8.5. Each part of the whole is a product of two halves.

In the case of dice, each of the 36 outcomes that comprise the table-type sample space (Fig. 8.3) is equally likely and, therefore, the fraction 1/36 is the measure of each outcome. At the same time, the likelihood of the event that after rolling two dice, the result is either eight or nine spots on both faces is measured by the fraction of the table filled with either eight or nine. As the two numbers appear nine times in the table, the probability of this event is 9/36 or 1/4. In that way, one can conclude that the chances of having head-head after tossing two coins are the same as having the sum of either eight or nine on both faces when rolling two dice. Apparently, without using fractions as measuring tools for chances (probabilities) this conclusion would not be possible.

One can also calculate the probability of the event that in four rolls of a fair die there appears no face with six spots. There are five chances out of six of not having six spots when rolling the die; that is, the probability is equal to 5/6. Therefore, because each roll of the die does not depend on the previous roll, the probability of not having a face with six spots over four rolls is equal to the product of four equal factors

$(5/6)(5/6)(5/6)(5/6) = (5/6)^4 \approx 0.482$. Likewise, one can measure the probability of a double six spots not appearing in a series of twenty-four rolls of two dice. As shown in the table of Fig. 8.3 (representing the sample space of rolling two dice), there are 35 chances out of 36 of not having a double six. That is, the probability of not having a double six in a series of twenty-four trials is equal to $(35/36)^{24} = 0.509$. The numbers 0.482 and 0.509 are somewhat close to each other being located by the different sides of 0.5 (which can serve as a benchmark number in ordering the other two). The numbers $(5/6)^4$ and $(35/36)^{24}$ will be used below to discuss a classic probability problem from the 17th century.

Note that in the above examples we talked about using fractions (or, more generally, numbers in the range [0, 1]) as tools in measuring theoretical probabilities of outcomes or events. As will be shown below, proper fractions can also be used in estimating experimental probabilities (alternatively, relative frequencies) of events as the ratio of the number of outcomes in favor of an event to the total number of trials used in the corresponding experiment. It is important to distinguish between the notions of probability and relative frequency because, as Mises [1957] put it, "a probability theory which does not introduce from the very beginning a connexion between probability and relative frequency is not able to contribute anything to the study of reality" (p. 63).

8.4 Bernoulli Trials and the Law of Large Numbers

An experiment (alternatively, a trial) of tossing a fair coin has only two outcomes, head or tail. When a coin is tossed two times, the outcome of the first toss has no influence on the outcome of the second toss. In other words, the two trials are independent and the probability of head or tail is the same for each experiment. Another example of a trial with two outcomes is the birth of a child; however, unlike tossing a coin, here the outcomes are not equally likely. As an aside, note that the study of the human sex ratio by Arbuthnot[20] [1710] nonetheless demonstrated that the excess of males over females is not due to chance; this study is

---

[20] John Arbuthnot (1667–1735) – a Scottish polymath, best known for his pioneering contributions to mathematical statistics due to his position as physician extraordinary to Queen Anne (1665–1714) – Queen of Great Britain and Ireland.

considered not only as one of the earliest investigations on this topic, but as the pioneering work on using a representative sample to make a generalization about a population (referred to as inferential statistics).

Despite the simplicity of the trials of that kind, they represent one of the most important concepts in the probability theory called Bernoulli[21] trials. Suppose that S_n is the number of heads observed within n tosses of a fair coin. Then S_n/n represents the fraction of heads in n tosses (trials). Bernoulli proved the following: the probability that the ratio S_n/n deviates from 1/2 by any number as small as one wants tends to the unity as n increases. This statement, of course, can be extended to any kind of Bernoulli trials with the probabilities of success and failure equal, respectively, to p and $q = 1 - p$ and it is known as the Law of Large Numbers in the form of Bernoulli.

Consider the following question related to the toss of a coin: Given the number of tosses, what is the likely length of the longest run of heads (or tails)? It turns out that there exists a formula allowing one to predict the likely length of the longest success run within a sequence of n Bernoulli trials. For example, according to this formula, within a real experiment one can predict that the likely length of the longest run of consecutive heads in 100 tosses is six. With this in mind, two groups of students may be asked to toss a coin 100 times, one group recording the results of a real experiment, another group recording the results of a simulated experiment. It appears that humans are not able to generate a random sequence through a simulated experiment [Reichenbach, 1949] and "almost all [simulated] experiments found systematic deviations from randomness" [Wagenaar, 1972].

In general, if p and q, $p + q = 1$, are the probabilities of success and failure within a single Bernoulli trial, then the largest length of consecutive successes within n Bernoulli trials (assuming nq is much larger than one) for which one could expect at least a single run of that length to occur is equal to the closest integer to $\log_{1/p}(nq)$ [Schilling, 2012]. In particular, when tossing a fair coin we have $p = q = 1/2$. In Fig. 8.6, the spreadsheet (which rounded $\log_2(10^i/2)$ to the closest

---

[21] Jacob Bernoulli (1654–1705) – a Swiss mathematician, a notable representative of the famous family of Swiss mathematicians.

integer value for $i = 1, 2, ..., 8$) shows the likely lengths of the longest run of consecutive heads (or tails) within the number of trials being the integer powers of ten. Similarly, other Bernoulli trials can be examined. For example, when rolling a die 500 times, the likely maximum length of consecutive sixes is 3 as in this case we have $p = 1/6$, $q = 5/6$ and the closest integer to $\log_6(500 \cdot 5/6)$ is 3. An interesting computer exploration is to verify data included in the chart of Fig. 8.6 by generating within a spreadsheet a sequence of, say, 1000 ones and zeroes to see the appearance of same numbers nine times in a row at least once, something that may not be observed within such a sequence of length 100. For that, one can use a spreadsheet formula =RANDBETWEEN(0, 1) replicated over 1000 (or, separately, 100) cells. This exploration can also be used as a test of the quality of a random numbers generator.

10	100	1000	10000	100000	1000000	10000000	100000000
3	6	9	13	16	19	23	26

Fig. 8.6. Rounding to the nearest integer the values of $\log_2(10^i/2)$ for $i = 1, 2, ..., 8$.

8.5 A Problem of Chevalier De Méré

The theory of probability as a disciplined inquiry evolved from the games of chance that were very popular in the 17th century France. One of such games was played as follows. A die was rolled four times in a sequel and one of the players bet to have six spots on the die to appear at least once; another player bet that this would not happen. A variation of this game was to use two dice and roll them twenty-four times; while one of the players bet to have a double six at least once, another player bet against it. History preserved the name of De Méré, an experienced gambler and erudite, who observed the following. In the game with one die, the chances of having at least one six in a series of four rolls were slightly higher than not having it; in the game with two dice, the chances of having at least one double six in a series of twenty-four rolls were slightly lower than not having it. De Méré believed that this discrepancy presents a paradox because, as his reasoning and computations might have gone, 4/6 = 24/36, where 4 and 24 are, respectively, the number of trials in the first and second games and 6 and 36 are the number of possible outcomes in each game. As De Méré was a gambler and not a mathematician, he, according to a number of sources (e.g., [Ore, 1960; Székely, 1986, p. 5]), asked Pascal, a compatriot and, most importantly, one of the founders of probability theory, to solve the problem, which ever since bears the gambler's name. In brief, Pascal's solution was based on the reasoning described above in section 8.3. The sample space of the game with one die consists of two events: having at least one six in a series of four rolls and not having it. Geometrically, the sample space represents one whole (the unity) comprised of two parts – one measured by $(5/6)^4$ and another one measured by $1-(5/6)^4$. The latter number is equal (approximately) to 0.517 and it represents the probability of having at least one six in four rolls of a die. The sample space of the game with two dice also consists of two events: having a double six at least once in a series of 24 rolls of two dice and not having it. Using the same reasoning as in the case of the first game, we have the probability of having a double six at least once in a series of 24 rolls equal $1-(35/36)^{24} \approx 0.491$. One can see that (long term) observations of De Méré were correct. In section 8.10.1 we will return to this problem and explain why his possible reasoning was nonetheless incorrect. In section 8.10.3 these two games will be simulated and the corresponding relative frequencies computed by using a spreadsheet.

8.6 A Modification of the Problem of De Méré

Suppose a fair die is rolled four times and one wants to compare chances of having six *at least* one time and *exactly* one time. The probabilities of having and not having six spots on a face of a die are 1/6 and 5/6, respectively. The tree diagram of Fig. 8.7 represents the sample space of this experiment with the letters S (success) and F (failure) used to indicate having and not having a six, respectively; the asterisks identify four rolls of a die with exactly one success of having a six. The tree shows that there are four (independent) outcomes in favor of a six appearing exactly one time in a series of four rolls of a die, with equal probabilities $(1/6)(5/6)^3$. The number of such (mutually independent) outcomes is equal to the number of permutations of letters in the word SFFF, that is, $C_4^1 = \dfrac{4!}{1!(4-1)!} = 4$ (Chapter 2, section 2.5). So, just as using the tree diagram of Fig. 8.2 allows one to conclude that the probability of having either HT or TH is the sum 1/4 + 1/4 (alternatively, $C_2^1(1/4)$), the tree diagram of Fig. 8.7 suggests that the probability of a six appearing exactly one time in a series of four rolls of a die can be calculated as follows: $C_4^1(1/6)^1(5/6)^{4-1} = \dfrac{4 \cdot 5^3}{6^4} \approx 0.386$. In general, the probability of exactly k successes in n Bernoulli trials with the probability of success p and the probability of failure q, $p + q = 1$, is equal to $C_n^k p^k q^{n-k}$. One can see that in a series of four rolls of a die, the probability of having a six *at least* one time (0.517) is higher than that of having a six *exactly* one time (0.386).

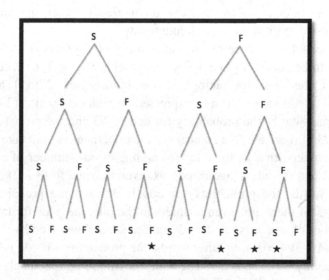

Fig. 8.7. Having a six *exactly* one time in a series of four rolls of a die.

Similarly, one can compare chances (probabilities) of having a double six at least one time and exactly one time in a series of twenty-four rolls of two dice. Recall, that the probabilities of having and not having a double six when rolling two dice are, 1/36 and 35/36, respectively. Therefore, the probability of the former event is equal to

$$C_{24}^1(1/36)^1(35/36)^{24-1} = \frac{24!}{1!(24-1)!}\frac{1}{36}\left(\frac{35}{36}\right)^{23} \approx 0.349.$$ Once again, in a

series of twenty-four rolls of two dice the chances of having a double six *at least* one time (0.491) are higher than that of having a double six *exactly* one time (0.349).

8.7 Wagering for the Odds/Evens in a Game of Chance

The ancient game of guessing odd or even number of objects in a hand (mentioned in the introduction to this chapter and that of Chapter 4) was notably explored through the lenses of probability theory in the 18th century. De Mairan[22] [1728] was the first to calculate the probabilities of odd and even number of objects in a hand using the following argument. Let objects be counters, our modern day manipulative materials. Because counters in a hand may only appear after they were randomly taken from a pile of counters, ‘one can make a distinction between the pile itself having an odd or even number of counters.

Consider the case of an odd number of counters in the pile. If there are three counters, then one can take either 1, or 2, or 3 counters. Therefore, the odds for taking an odd number are 2 to 1 and the advantage of odd vs. even can be expressed through the fraction $(2 - 1)/2 = 1/2$. In other words, the probability for odd is 2/3 and the probability for even is 1/3. If there are five counters in the pile, then one can take 1, 2, 3, 4, or 5 counters; that is, the odds for taking an odd number of counters are 3 to 2 and the advantage of odd vs. even is equal to $(3 - 2)/3 = 1/3$. In other words, the probability for odd is 3/5 and the probability for even is 2/5. If there are seven counters in the pile, the odds for taking an odd number of counters are 4 to 3 and the advantage of odd vs. even is equal to $(4 - 3)/4 = 1/4$. In other words, the probability for odd is 4/7 and the probability for even is 3/7. The table shown in Fig. 8.8. displays more cases. However, when the number of counters in the pile is even (Fig. 8.10), there is no advantage of odd vs. even and the probabilities of having an odd and even number of counters in a hand are the same.

But what is even more interesting about this game and its probabilistic description is that the above solution by de Mairan was criticized by such a great mind as Laplace[23] [1774] in the following

---

[22] Jean-Jacquest d'Ortous de Mairan (1678–1771) – a French scientist.

[23] Pierre Simone Laplace (1749–1827) – a French mathematician and scientist, one of the greatest scholars of all time.

words: "Mr. de Mairan has likewise observed that there is always a greater advantage to wager for the odds than for the evens; but it seems to me that the manner in which this ingenious author considers the problem is not correct" (p. 13). Unfortunately, the solution by Laplace is beyond the elementary level and it is only possible to present numerically the probabilities computed by Laplace [1774] through a spreadsheet pictured in Fig. 8.11. One can see (Fig. 8.9 and Fig. 8.12) that regardless of a method used in calculating probabilities (or, better, relative frequencies) of an odd number of counters in a hand, the frequencies tend to 1/2.

odd pile of counters	3	5	7	9	11	13	15	17	19	21
odd counters	2	3	4	5	6	7	8	9	10	11
even counters	1	2	3	4	5	6	7	8	9	10
advantage	1/2	1/3	1/4	1/5	1/6	1/7	1/8	1/9	1/10	1/11
Prob. odd	0.667	0.6	0.571	0.556	0.545	0.538	0.533	0.529	0.526	0.524
Prob. Even	0.333	0.4	0.429	0.444	0.455	0.462	0.467	0.471	0.474	0.476

Fig. 8.8. Solution by de Mairan in the case of an odd pile.

Fig. 8.9. The behavior of relative frequencies of odds according to de Mairan.

even pile of counters	2	4	6	8	10	12	14	16	18	20
odd counters	1	2	3	4	5	6	7	8	9	10
even counters	1	2	3	4	5	6	7	8	9	10
advantage	0	0	0	0	0	0	0	0	0	0

Fig. 8.10. Solution by de Mairan in the case of an even pile.

pile of counters	2	3	4	5	6	7	8	9	10	11	12
prob. for odd	0.667	0.571	0.533	0.516	0.508	0.504	0.502	0.501	0.5	0.5	0.5
prob. for even	0.333	0.429	0.467	0.484	0.492	0.496	0.498	0.499	0.5	0.5	0.5

Fig. 8.11. Solution by Laplace.

Fig. 8.12. The behavior of relative frequencies of odds according to Laplace.

8.8 Paradoxes in the Theory of Probability

8.8.1 *Bertrand's Paradox Box problem*

Using the concept of sample space can be helpful in explaining probabilistic reasoning the result of which may be perceived as counterintuitive. To this end, consider the famous Bertrand's Box Paradox[24]. There are three boxes, each filled with two coins. The first box (GG) is filled with two gold coins, the second box (GS) is filled with one gold coin and one silver coin, and the third box (SS) is filled with two silver coins. The question to be answered is: *Assuming that a box is chosen at random from which a gold coin is drawn without looking, what is the probability that the second coin also randomly drawn would be a gold coin as well?* Intuitively, it appears that the probability for the second coin, when drawn without looking, to be also a gold coin is equal to 1/2. Indeed, it seems to be equally likely to pick up a gold coin second time either from the GG box or from the GS box. However, the right answer is 2/3.

In order to explain how this answer can be obtained, let us recall that the sample space of tossing two coins, {HH, HT, TH, TT}, consists of four outcomes with known probabilities each of which is a product of

---

[24] Joseph Louis François Bertrand (1822–1900) – a French mathematician.

two one-halves, because the probability of head or tail for a single coin is 1/2. It is with certainty, that one of the four outcomes with probability 1/4 would take place when tossing two coins. Therefore, because geometrically, the sample space represents one whole (the unity), the sum of the probabilities of the four outcomes is equal to one. In the case of the Bertrand's Paradox Box the sample space of selecting a gold coin twice can be represented in the form $\{GG/G_{GG}, GG/G_{GS}, GG/G_{SS}\}$. The fractional notation used in representing the sample space means the following. The numerator GG is the outcome of selecting a gold coin twice, the probability $P(GG)$ of which is unknown (and has to be found); each of the three denominators represent, respectively, the following outcomes related to the first selection of a gold coin: the coin belongs to the GG box (with probability $P(G_{GG}) = 1$); the coin belongs to the GS box (with probability $P(G_{GS}) = 1/2$); the coin belongs to the SS box (with probability $P(G_{SS}) = 0$). By analogy with the case of tossing two coins, we have $P(GG) \, P(G_{GG}) + P(GG) \, P(G_{GS}) + P(GG) \, P(G_{SS}) = P(GG)(1 + 1/2 + 0) = P(GG)(3/2) = 1$, whence $P(GG) = 2/3$.

8.8.2 *Monty Hall Dilemma*

Monty Hall was a popular host of an American television game show of the 1970s. His name is associated with the following paradox of the probability theory that has occurred during the following game show. Behind three closed doors an expensive car and two whimsical prizes are hidden and a contestant in the show must choose between the doors aiming at winning the car. He or she chooses a door at random (but does not open it), thereby wins the car with a 1/3 probability. Then, the host, who knows where the car is, opens one of the other two doors with a whimsical prize behind it. Now, with one door open and two doors closed, the host asks the contestant to make the next choice: either to remain with the original selection of the door (thus winning the car with a 1/2 probability) or to switch to another (closed) door with a possibility to increase the chances of winning the car. This quandary is called Monty Hall Dilemma [vos Savant, 1996].

In 1990, an American columnist Marilyn vos Savant, in her weekly *Parade* column, advised the contestant to switch to the other door because this switch would raise the odds of winning the car to a 2/3 probability. Many readers of the column, professional mathematicians among them, strongly objected (see [vos Savant, 1996] for details) to the claim that switching the choice of a door would increase the probability of winning the car from 1/2 to 2/3. Yet, vos Savant's advice was correct. To explain, note that whereas the contestant's original choice of a door was random, the host's choice of the door was not random, informed by the knowledge of the location of the car. This added additional value to the door not selected by the host.

A reasoning that leads to the answer 2/3 can be demonstrated through the use of a tree diagram shown in Fig. 8.13. Let D1, D2, and D3 be the three doors. The contestant can choose one of the doors with a 1/3 probability selecting a door with the car behind it. If the contestant chooses D1 and it conceals the car, then the host opens either D2 or D3 with a 1/2 probability. If the door D1 (chosen by the contestant) does not conceal the car, but D2 does, then the host would open D3 with the probability one. Likewise, if the door D1 (chosen by the contestant) does not conceal the car, but D3 does, then the host would open D2 with the probability one. Now, the probabilities for each situation can be computed. Just as in the case of tossing two coins when the probability of each of the four outcomes is the product of the probabilities of a single outcome, the probabilities of the two actors' choices in the case of Monty Hall Dilemma have to be multiplied. Therefore, if door D1 is chosen by the contestant and it conceals the car, then, whatever the door the host opens, the probability of winning the car without switching doors is equal to $\frac{1}{3} \cdot \frac{1}{2} + \frac{1}{3} \cdot \frac{1}{2} = \frac{1}{3}$. However, if D1 conceals the car, but the contestant first chooses D2 or D3, the decision of switching from D2 to D1 (with D3 open) or from D3 to D1 (with D2 open) yields the following probability of winning the car: $\frac{1}{3} \cdot 1 + \frac{1}{3} \cdot 1 = \frac{2}{3}$. This shows the correctness of vos Savant's advice and resolves the Monty Hall Dilemma.

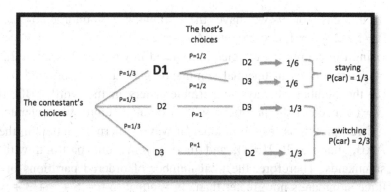

Fig. 8.13. Resolving Monty Hall Dilemma.

8.9 Probabilistic Perspective on Partitioning Problems

8.9.1 *A problem of tossing three dice*

In Chapter 1, section 1.2, a possible query into partitioning a natural number into a sum of more than two like numbers was mentioned and in Chapter 5, section 5.5, the values of $P(n,m)$ – the number of unordered partitions of n into m summands – was generated for $1 \le m \le n \le 11$ by a spreadsheet (Fig. 5.10). In particular, $P(10,3) = 8$ and $P(9,3) = 7$. In this context, one can mention a famous problem solved by Galileo[25] who was asked by a friend [Todhunter, 1949, pp. 4–5] as to why when rolling three dice the number 10 appears more often than the number 9, although each can be partitioned in three unordered positive integer summands not greater than six in six ways. According to Tijms [2012], this problem was posed to Galileo not by a friend but rather by "the Grand Duke of Tuscany, his benefactor" (p. 2).

Nowadays, this problem can be easily solved by using a combination of methods discussed in this book. To this end, note that the equality $P(9,3) = 7$ includes the partition $9 = 1 + 1 + 7$ and the equality $P(10,3) = 8$ includes two partitions, $10 = 1 + 1 + 8$ and $10 = 1 + 2 + 7$, something that is not possible to observe when rolling a six-sided die. So,

---

[25] Galileo Galilei (1564–1642) – an Italian scholar, the father of the major scientific developments of the 17th century.

in the case of 9 we have the following six unordered partitions: $9 = 1 + 2 + 6 = 1 + 3 + 5 = 1 + 4 + 4 = 2 + 2 + 5 = 2 + 3 + 4 = 3 + 3 + 3$. Using the counting strategies introduced in Chapter 2, section 2.3, each of the partitions in unequal summands generates six ordered partitions (as the number of ways to permute letters in the word ABC, that is, $3! = 6$); each of the partitions with two equal summands generates three ordered partitions (as the number of ways to permute letters in the word AAB, that is, $3!/2! = 3$); and there is only one partition with equal summands. Therefore, the total number of ordered partitions of 9 into three summands not greater than six is equal to $6 + 6 + 3 + 3 + 6 + 1 = 25$ (counted in the order presented in the above list of unordered partitions of 9). Likewise, in the case of 10 we have $10 = 1 + 3 + 6 = 1 + 4 + 5 = 2 + 2 + 6 = 2 + 3 + 5 = 2 + 4 + 4 = 3 + 3 + 4$ and the number of ordered partitions of 10 into three summands not greater than six is equal to $6 + 6 + 3 + 6 + 3 + 3 = 27$. Regardless who posed the problem to Galileo – a friend or the Grand Duke of Tuscany, Galileo found that the probabilities of appearing 10 and 9 when rolling three dice are, respectively, 27/216 and 25/216, where $216 = 6^3$ is the total number of sums appearing on three six-sided dice.

8.9.2 *Unordered partitions of integers into unequal summands*

One can revisit other partitioning problems discussed in Chapter 5 from a probabilistic perspective where the problems were motivated by activities of building towers under specified conditions using square tiles. These conditions, along with the number of tiles used, determine the sample space of the experiment. For example, as shown in Chapter 5 through the spreadsheet of Fig. 5.13 and confirmed at the level of the first-order symbolism in Fig. 5.14, there exist five unordered partitions of the number 14 into four unequal summands; in other words, one can construct five sets of four different size towers out of 14 square tiles. The sample space of the experiment of selecting at random a set of towers of that kind is shown in Fig. 5.14 using the first-order symbols. At the level of the second-order symbolism, the sample space can be represented in the form of a set of quadruples of integers

$$\{(1, 2, 3, 8), (1, 2, 4, 7), (1, 2, 5, 6), (1, 3, 4, 6), (2, 3, 4, 5)\}.$$

Alternatively, without explicitly constructing the sample space, one can use formula (5.4) of Chapter 5 to recursively compute the number of requires partitions

$$Q(14, 4) = Q(10,3) + Q(10,4) = Q(7,2) + Q(7,3) + Q(6,3) + Q(6,4)$$
$$= 3 + 1 + 1 + 0 = 5.$$

Consequently, one can be asked to calculate the probability that the unordered partition of 14 into four unequal summands consists of consecutive natural numbers. Knowing that $Q(14, 4) = 5$, does not guarantee that the number 14 has a trapezoidal representation. To find such representation, one has to represent 14 in the form $7 \cdot 4 / 2$, which points at a trapezoidal representation with four rows. Finally, noting that $15 = 1 + 2 + 3 + 4 + 5$ yields $14 = 2 + 3 + 4 + 5$. Thus the outcome sought exists and, assuming that it is equally likely to select any unordered partition of 14 into four unequal summands, its probability is equal to 1/5.

8.10 Experimental Probability

8.10.1 *Experimental probability calls for a long series of observations*
Experimental probability (relative frequency) of an event is defined as the ratio of the number of times the event occurred in a series of identical trials to the total number of trials in this series. As the number of trials grows larger, the difference between experimental and theoretical probabilities becomes smaller. That is, relative frequency can be used to replace theoretical probability when the number of trials is sufficiently large (*cf.* the Law of Large Numbers, section 8.4). This relationship between experimental and theoretical probabilities is especially useful didactically when theoretical probability is difficult to determine, either from a pure mathematical perspective or because such determination requires reasoning tools of formal mathematics that are far beyond its grade-appropriate pedagogy. This is when experimental reasoning techniques can be employed to replace probability by relative frequency. In the case of the problems by De Méré, while his true observations were based on a large number of rolls of dice, a possible reasoning about equal chances of having and not having at least one six or at least one double six was based on only four and twenty four tosses, respectively. In the

several millenniums span, people who played the games of chance learned how to make a fair die (its first appearance dates back to the 3rd millennium BC [David, 1970; Bennett, 1998]). Through playing such games, it was found that the chances of having and not having at least one six in a series of four rolls of a die were about the same. Likewise, the chances of having and not having a double six at least one time in a series of twenty-four rolls of two dice were about the same. But in a long practice of playing the games it was observed that chances deviated[26] from 50%. The problems of De Méré thereby show "the empirical character of the theory of probability and its purpose of interpreting observable phenomena" [Mises, 1957, p. 64].

8.10.2 *Comparing experimental and theoretical probabilities when tossing a fair coin*

The importance of using experimental techniques in the study of probability theory at the pre-college level is acknowledged by many standards worldwide. For example, students in Canada are expected to "pose and solve simple probability problems, and solve them by conducting probability experiments" [Ontario Ministry of Education, 2005, p. 85]. In the United States, students "learn about the importance of representative samples for drawing inferences" [Common Core State Standards, 2010, p. 46]. In England, expectations for students in the context of probability include the ability to "record, describe and analyse the frequency of outcomes of simple probability experiments involving randomness" [Department for Education, 2013, p. 42]. Probability experiments can be facilitated by the use of technology, in particular, by using a spreadsheet.

Consider the case of tossing a fair coin and recording the frequency of heads (or tails) in this experiment. The theoretical probability of having a head (or a tail) when tossing a fair coin is equal to 1/2. Experimentally, using different simulation techniques one can show, when tossing a coin a large number of times, that the frequencies of heads and tails would be close to each other. Fig. 8.14 shows the use of a spreadsheet in experimentally calculating the probability of a coin turning up head when each of the 20 students in a class tossed a coin 100

---

[26] The nature of these deviations was explained in section 8.3.

times. It turned out that among 2000 tosses, the frequency of heads was 994 yielding the relative frequency of heads in this experiment equal to 0.497 – a number close to 1/2.

Alternatively, taking advantage of the tool's feature to generate random numbers within a specified range, this experiment can be made purely digital: the appearance of the number 1 in column C (or D) indicates that a coin turned up head (or tail). The spreadsheet of Fig. 8.15 simulates 2000 tosses of a coin, repeats this experiment 500 times (cell F1) and displays the experimental probability (relative frequency) of heads in cell E3 by averaging the relative frequencies of heads (displayed in column E) for each of the 500 experiments. (Cells C2 and D2 display relative frequencies of head and tail, respectively, within the last experiment). One can see (cell E3) that the relative frequency of a coin turning up head out of 1000000 tosses, being approximately equal to 0.4996, is closer to the theoretical probability even in comparison with 2000 tosses (0.497) let alone a small number of tosses. One can check to see that tossing a coin only ten times may give an experimental result significantly different from the theoretical one. This illustrates the Law of Large Numbers (section 8.4) and the importance of dealing with relative frequencies in long series of identical trials.

		Exp. Pr. (Heads)	Exp. Pr. (Tails)	
		0.497	0.503	
				Heads total
	Name	Heads	Tails	994
1	Jody	47	53	
2	Rachel	45	55	
3	Sarah	52	48	
4	Ellen	53	47	
5	Laura	50	50	
6	Joshua	46	54	
7	Jenn	55	45	
8	Latisha	53	47	
9	Lynn	60	40	
10	Holly	49	51	
11	Beth	44	56	
12	Derek	47	53	
13	Cathy	52	48	
14	Nora	55	45	
15	Amy	51	49	
16	James	45	55	
17	Katie	43	57	
18	Jon	54	46	
19	Molly	49	51	
20	Teresa	44	56	
21				

Fig. 8.14. Recording 100 tosses of a coin by each of 20 students.

	A	B	C	D	E	F
1			Exp. Pr. (Heads	Exp. Pr. (Tails)	Heads Total	500
2			0.511	0.489	1022	
3					0.499644	
4			Head	Tail		
5			0	1	0.4925	
6			1	0	0.5035	
7			0	1	0.4865	
8			1	0	0.5075	
9			1	0	0.4945	
10			0	1	0.5025	
11			0	1	0.516	
12			0	1	0.4865	
13			1	0	0.4925	
14			1	0	0.485	
15			1	0	0.5	
16			0	1	0.4955	
17			1	0	0.516	
18			0	1	0.4915	
19			1	0	0.4845	
20			0	1	0.5035	
21			0	1	0.5155	
22			0	1	0.49	
23			0	1	0.502	
24			1	0	0.4995	
2004			0	1		

Fig. 8.15. Spreadsheet simulation of 1000000 tosses of a coin.

8.10.3 *Calculating relative frequencies for the problems of De Méré*

A spreadsheet has a feature of generating random numbers within a specified range. For example, in section 8.4 a spreadsheet was suggested as a generator of long random series of zeroes and ones in order to observe the longest length of identical digits. Once a random sequence of numbers has been generated, one can explore chances that the numbers or their combinations possess a certain property by calculating relative frequencies of certain properties to be observed (called attributes). As mentioned by Mises [1957], it is important to demonstrate that "the relative frequencies of certain attributes become more and more stable as the number of observations is increased" (p. 12).

The spreadsheet of Fig. 8.16 simulates a long sequence of four rolls of a die and records the total number of appearances of at least one six within each series of four rolls. Each click at the spin button attached to cell F1 initiates four rolls of a die and automatically records the appearance of six spots. The number of such clicks is displayed in cell F1 and the total number of successes (that is, observing the attribute "6") is recorded in cell E1. The spreadsheet pictured in Fig. 8.16 shows the final result – after 500 series of four rolls of a die at least one six was observed 262 times. That is, (within this experiment) the relative frequency of having at least one six in a series of four rolls equals 262/500 = 0.524, something that coincides with the observations of De Méré.

	A	B	C	D	E	F	G
1					262	500	
2	2			1			
3	4			2		▲	
4	3			3		▼	
5	3			4			
6				5	1		
7				6			
8				7	1		
9				8	1		
10				9	1		
11				10	1		
12				11			
13				12			
14				13	1		

Fig. 8.16. Simulating 4 rolls of a die 500 times.

	A	B	C	D	E	F	G
1					243	500	
2	3	2		1	1	▲	
3	5	2		2	1	▼	
4	1	5		3			
5	6	5		4	1		
6	6	3		5	1		
7	1	5		6			
8	5	6		7	1		
9	2	3		8			
10	2	5		9			
11	2	4		10			
12	1	6		11	1		
13	4	2		12			
14	5	4		13			
15	3	3		14	1		
16	6	1		15	1		
17	4	3		16	1		
18	2	4		17	1		
19	3	1		18	1		
20	5	3		19			
21	2	6		20			
22	4	6		21	1		
23	4	5		22	1		
24	1	4		23	1		
25	6	2		24	1		
26				25			

Fig. 8.17. Simulating 24 rolls of two dice 500 times.

Similarly, the spreadsheet (Fig. 8.17) that simulates the second game by De Méré shows that the relative frequency of having at least a double six in a series of twenty-four rolls of two dice equals 243/500 = 0.486. Once again, the computational experiment confirms the gambling observations.

8.11   Exploring Irreducibility of Fractions through the Lenses of Probability

In this section, fractions, which, at the primary school level, can be presented as tools in computing chances of simple events, will themselves be considered through the lenses of probability. To this end, consider the problem of determining probability of co-primality of two randomly selected natural numbers (that is, the probability that their greatest common divisor is equal to one). When such pairs of numbers (also called relatively prime numbers) are used as the numerator and the denominator of a fraction, it is known of being presented in the simplest form; in other words, the fraction is irreducible. In such a way, the problem may be given the following equivalent formulation: find the probability that two randomly selected natural numbers form an irreducible fraction. The theoretical probability of such an event, easy to understand but difficult to examine, is equal to $6/\pi^2 \approx 0.608$ (where π is the ratio of circumference to its diameter).

Unfortunately, the proof of this (purely theoretical) result requires means far beyond the elementary level. Notwithstanding, especially in the digital age, "the fact that proof is important for the professional mathematician does not imply that the teaching of mathematics to a given audience must be limited to ideas whose proofs are accessible to that audience" [Stewart, 1990, p. 187]. Furthermore, this is a classic problem, with a truly rich history and, according to the Conference Board of the Mathematical Sciences [2012], historical topics when integrated with "the proper use of technology [can] make complex ideas tractable" (p. 67). The problem may safely be considered among the most famous problems in the history of mathematics for it is associated with the names of its distinguished creators. Most likely it was posed and solved by Dirichlet[27] and, by some accounts, was known to Gauss[28] [Arnold, 2015] and even to Euler [Arnold, 2005]. The problem and its solution can also be found in the modern day mathematics education journals [Abrams and Paris, 1992; Hombas, 2013].

In what follows, the problem will be approached experimentally by using a spreadsheet. Such spreadsheet is shown in Fig. 8.18 where 10000 pairs of randomly selected natural numbers not greater than 10000 have been generated in columns A and B. In column C their greatest common divisor is computed and the number of cells filled with the unity is computed in cell D2. The spreadsheet calculates experimental probability of the relative primality/irreducibility out of 10000 trials 500 times (cell G2), records the results in column F, and then takes the average of 500 experiments (cell G4). In that way, the total number of trials is equal to $5 \cdot 10^7$.

The spreadsheet of Fig. 8.18 is programmed as follows. In columns A and B, using the function =RANDBETWEEN(1, 10000), the spreadsheet generated 10000 pairs of natural numbers. In column C the spreadsheet tests the pairs through the greatest common divisor function and displays the values of this function for 10000 consecutive pairs. For example, as shown in Fig. 8.18, GCD(9582, 91) = 1 (cell C3) and GCD(6342, 6954) = 6 (cell C2). That is, the first two numbers are relatively prime and the second two numbers are not. A special technique (not discussed here) makes it possible to carry out this experiment more than one time (500 times as shown in the spreadsheet of Fig. 8.18) and then take the average value of 500 so computed experimental probabilities. This value, 0.60812, is displayed in cell G4 and compared to the theoretical probability, $6/\pi^2$, (cell I2) with the error 0.0002 displayed in cell K2. The behavior of relative frequencies for the first 20 experiments is shown in Fig. 8.19 demonstrating what the theory predicts: the graph of relative frequencies, by oscillating, gradually approaches the level of 0.608.

---

[27] Peter Gustav Lejeune Dirichlet (1805–1859) – an outstanding German mathematician.

[28] Carl Friedrich Gauss (1777–1855) – a German mathematician, commonly regarded as the greatest mathematician of all time.

	A	B	C	D	E	F	G	H	I	J	K
1	a	b	GCD(a, b)	GCD(a, b)=1		Exp. Probability			Th. Probability		error
2	6342	6954	6	6048	1	0.603360336	500		0.607927102		0.000207912
3	9582	91	1		2	0.602960296					
4	8789	245	1		3	0.614461446	0.60812				
5	2010	7359	3		4	0.609460946					
6	419	6124	1		5	0.608260826					
7	629	1125	1		6	0.604260426					
8	7668	1508	4		7	0.611761176					
9	6246	1387	1		8	0.616561656					
10	2630	7481	1		9	0.601560156					
11	3780	3793	1		10	0.599259926					
12	7731	5100	3		11	0.613961396					
13	3304	3922	2		12	0.604060406					
14	3432	9039	3		13	0.603060306					
15	3046	2191	1		14	0.609460946					
16	9940	3969	7		15	0.608160816					
17	3236	4289	1		16	0.609460946					
18	2672	3293	1		17	0.612661266					
19	383	5955	1		18	0.603960396					
20	5057	2670	1		19	0.607260726					
21	3326	329	1		20	0.602860286					
9999	9577	4141	1								
10000	4690	8650	10								

Fig. 8.18. Using a spreadsheet in calculating the probability of irreducibility of a fraction.

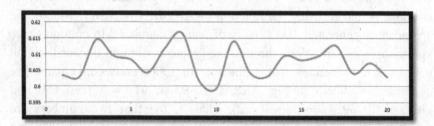

Fig. 8.19. The behavior of relative frequencies calculated by a spreadsheet.

Chapter 9

Using Counter-Examples in the Teaching of Elementary Mathematics

9.1 Introduction

This chapter deals with the use of counter-examples as one of the methods of teaching mathematics. The literature on counter-examples (sometimes referred to as non-examples) is mostly devoted to the tertiary [Steen and Seebach, 1995; Mason and Klymchuk, 2009; Klymchuk, 2010] or the secondary [De Villiers, 2004; Peled and Zaslavsky, 1997; Zazkis and Chernoff, 2008] levels of mathematics curricula. Nonetheless, counter-examples have been reported as valuable means of support of mathematics teaching at the primary level as well [e.g., Tirosh and Graeber, 1989]. How can one explain to a second grader that, unlike addition, the operation of subtraction is not commutative? One may ask whether it is possible to buy a $2 ice cream having $1 only. The negative answer to this question is contextually pretty obvious and its appropriate de-contextualization using the second-order symbolism can demonstrate the difference between addition and subtraction in this regard. It also shows that a general statement like 'with $1, I can buy *anything*' is not true. Yet, the ice cream counter-example not only demonstrates that subtraction is not commutative, but it can also be used to motivate the extension of the system of natural numbers to give meaning to subtraction beyond the primary school constraint of having the minuend not smaller than the subtrahend.

The use of counter-examples in mathematics education has three major goals. The first goal is to help learners to conceptualize a certain mathematical property, like non-commutativity of the operations subtraction and division. The second goal is to demonstrate that certain mathematical statement lacks generality, such as the assertion that all odd numbers are primes. The third goal is to avoid the so-called Einstellung effect (Chapter 7, section 7.4, Remark 7.4) – an uncritical application of a mastered problem-solving strategy (method) as one moves from a seemingly grasped part of the curricula (e.g., the preservation of the sign of an inequality between two positive numbers by doubling them) to a new conceptual domain (involving negative numbers). One can see that the use of counter-examples as a mathematics teaching method requires robust knowledge of the subject matter. Put another way, in order to develop a pedagogical skill of using counter-examples one has to possess a skill of navigating within the whole elementary mathematics content and to appreciate both procedural and conceptual connections among different parts of the content. In this final chapter, reflecting on the material of the preceding chapters, several grade-appropriate examples supporting the above three goals and boosting the skill of the corresponding pedagogy will be presented.

9.2   The Pedagogy of Using Counter-Examples

The three goals of using counter-examples in the preparation of elementary teacher candidates also serve at least two didactic purposes. The first purpose is to explain why certain mathematical action has to be taken or a particular linguistic form has to be used when formulating a task. Prospective teachers of primary school mathematics may lack mathematical sophistication expected from their secondary level counterparts. Something that might seem obvious to the latter group could be not so obvious to the former group. In many cases, the natural curiosity of learners encouraged by current standards for teaching mathematics at the primary level [e.g., Common Core State Standards, 2010] can be a source for unexpected questions, the finding of answers to which might be pedagogically challenging.

9.2.1   *The role of linguistic constraints*

As described in Chapter 3 (section 3.2.3), when a teacher candidate during her internship asked kindergarteners which color do they have the most/least among the handful of square tiles that they were given, it is only by providing a counter-example of having same number of the tiles in two colors that one can explain why the task should be formulated slightly different by adding the word *same*. For example, young children could be asked: Is there a color that you have the most and/or the least and are there colors that you have the same? Without using a counter-example, the linguistic meaning of the text of a mathematical task may be simply unclear to inexperienced teacher candidates. Similarly, from the measurement perspective, the concept of perimeter is easier than the concept of area. Frequently, when asked how to find perimeter and area of a rectangular shape (e.g., a desk or a whiteboard), teacher candidates offer an action-like answer (measure distance around the desk) for perimeter and formula-like answer (multiply length by width) in the case of area. Perhaps, this dichotomy of responses is because there are common tools to measure distance (e.g., a measuring tape), but no obvious tools to measure area. A tool for measuring area has to be conceptualized and it is a unit square. Experiments carried out in "a fifth grade class of 36 pupils in a public school in a Brooklyn [New York] slum" [Luchins and Luchins, 1970a, p. 261] indicate that after such conceptualization was acquired, "the entire class knew how to determine the number of squares in a row by measuring ... paper rectangles [with a ruler] ... [and thereby, could also find] the areas of their desks, the classroom's floor, and the blackboard" [*ibid*, p. 262].

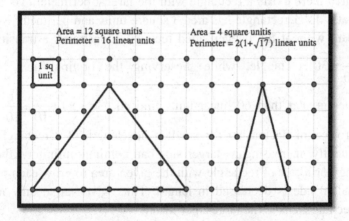

Fig. 9.1. Perimeter of the smaller triangle is not a rational number.

The use of a geoboard is helpful in demonstrating two different units of measurement. However, although area is conceptually (and, to some extent, computationally) more difficult than perimeter, at the elementary level, only areas (not perimeters) of polygons can be explored on a geoboard. With this in mind, it has to be explained to teacher candidates that whereas on a geoboard, the area *A* of any not semi-intersecting polygon is a multiple of one-half, something that Pick's formula, $A = \dfrac{B}{2} + I - 1$, expressing area through the number of pegs B and I, respectively, on the polygon's border and in its interior confirms (Chapter 7), already perimeter of such a simple polygon as triangle can be expressed through a rational (in fact, natural) number in some special cases only (Fig. 9.1). To explain this difference between perimeter and area on a geoboard, one has to be reminded about Pythagorean triples that bring into existence the above mentioned special cases (by themselves serving as counter-examples to the statement that perimeter of a triangle on a geoboard may not be a rational number). For example, the triple (3, 4, 5) represents the side lengths of a right-angled triangle (with perimeter 12) which can be constructed on a geoboard[29].

Another example of the importance of linguistic constraints in mathematical formulations of a geometric task is the reference to whole number sides when making a statement that among all rectangles with given area there exists a rectangle with the largest perimeter. To clarify, note that a 1×3 rectangle has area 3 square units and perimeter 8 linear units. But without the integer sided rectangle constraint, a transition to the $\dfrac{1}{10} \times 30$ rectangle, while preserving the original area, makes perimeter greater than 60 linear units (indeed, $30 + 30 + \dfrac{1}{10} + \dfrac{1}{10} > 60$). This process of decreasing the smaller side length of a rectangle and, consequently, increasing its larger side can continue infinitely, thereby, enabling perimeter of rectangle with the given area to grow each time as the smaller side is decreased in length. Thus, given area, only integer sided rectangles have the largest perimeter.

---

[29] As an aside, it appears interesting to construct on a geoboard other types of polygons (different from rectangles) with integer perimeters and explore their properties.

9.2.2 *A counter-example as a motivation for further learning*

The second pedagogical purpose of using counter-examples in an elementary mathematics education content and methods course is to demonstrate how to motivate young children to ask questions and how to survive a possible complexity of the questions by answering them appropriately. At the very basic level, whereas one can use counting to find the number of apples in two bags with two and three apples, respectively, in the case of the bags filled with hundreds of apples, the counting strategy becomes ineffective. That is, the increase in the magnitude of numbers can serve as a counter-example to mistakenly perceived efficiency of counting at the pre-operational level. This provides an answer to a question about the purpose of addition as an operation.

At a more intricate, yet still elementary, level, as shown, for example, in Chapter 3, section 3.10, second graders were able to find all ways to put five rings on two fingers experientially and, due to their natural curiosity, asked for an answer in the case of five rings and three fingers. This request for information, not easily available through an experiment and, instead, requiring the use of the "Y/B" model (Chapter 2, section 2.6), may be interpreted as a counter-example demonstrating that hands-on counting techniques of elementary combinatorics very soon reach their limitation. That is, a natural move from two fingers to three fingers can be used as a counter-example to naïve perception that one can do challenging mathematics at the level of the first-order symbolism only. Furthermore, this counter-example shows one of the most characteristic features of mathematics – a slight modification of a mathematical condition may lead to a situation that even an experienced teacher may not know how to resolve. This feature of mathematics has to be communicated to teacher candidates within an elementary mathematics content and methods course as appropriate.

9.3 Providing Explanation through Counter-Examples

Another type of counter-examples oriented towards asking questions may serve as the demonstration of a difference between questions seeking information and questions requesting explanation [Isaacs, 1930]. The goal of such demonstration is to show that the former questions are much more difficult to answer than the latter questions. Returning to the above example with rings, there are six ways to put five rings on two fingers and twenty-one ways to put them on three fingers[30]. A question seeking explanation of this quantitative information might be why adding only one finger leads to the increase in the number of ways to wear the rings by fifteen. One can note that $15 = 1 + 2 + 3 + 4 + 5$. Can the increase by 15 be explained in terms of the first-order symbols and then described at the level of the second-order symbolism as the sum of the first five natural numbers? In answering this question, one can introduce (or use once again) the method of reduction to a simpler problem; that is, the reduction from a problem with three fingers to a problem with two fingers. To this end, one can note that the third finger may have a number of rings in the range zero through five. The zero case may be immediately left out because it is equivalent to putting all the rings on two fingers. The case of five rings on the third finger renders one way, the case of four rings on the third finger renders two ways, the case of three rings – three ways, the case of two rings – four ways, and the case of one ring – five ways. All the cases can be confirmed in terms of the first-order symbols and then described through the second-order symbolism in terms of the sum $1 + 2 + 3 + 4 + 5 = 15$.

In connection with the last equality, one can ask whether any integer can be represented as a sum of consecutive integers (not necessarily starting from one)? This question seeks generality of a certain property and there are two ways an answer can be developed: either to prove generality or to disprove it by providing a counter-example. As shown in Chapter 6, section 6.9 (Remark 6.3), in order for such a representation of an integer N to exist, the double of N must be a product of two natural numbers of different parity, something that is not possible for a power of two. This proves that the statement of generality regarding representation of integers as a sum of consecutive natural numbers is not true. However, whereas the remark was deducted from a series of propositions of different levels of complexity, its statement can be proved through a counter-example using, say, number 4, which,

---

[30] Indeed, using the "Y/B" model (Chapter 2, section 2.6), the number 21 can be found as the number of permutations in the word YYYYYBB (five rings and two spaces among three fingers): $7!/(5! \cdot 2!) = 21$.

obviously, does not have such a representation. Therefore, a counter-example in some cases provides an easy demonstration that a certain mathematical property lacks generality.

9.4 Constructing a Counter-Example: an Illustration

9.4.1 *From modeling with fractions to algebraic generalization*
Consider the following (real-life) situation.

>   *Alan and Ana like to eat red M&Ms only. There are two bags of M&Ms available, plain and peanut. There are 3 red out of 5 total in the bag with plain M&Ms and 4 red out of 7 total in the bag with peanut M&Ms. How many red M&Ms does Alan have to eat from each bag so that Ana, without looking, has more chances to pick up a red peanut M&M than red plain M&M.*

The situation can be resolved through comparing pairs of fractions that represent chances numerically. At the beginning, the chances for a red M&M for each bag are expressed by the fractions 3/5 and 4/7 which are in the relation 3/5 > 4/7 (see Chapter 4, section 4.4.1 for the use of the two-dimensional method of comparison of the two fractions). After Alan eats one red M&M from each bag, the chances for a red candy to be picked up become the same for each bag, due to the equality 2/4 = 3/6. But after Alan ate one red M&M from each bag again, the chances become represented by the fractions 1/3 and 2/5 satisfying the inequality 1/3 < 2/5. The last relation between the chances implies that now, for the first time, Ana has more chances to get a red peanut M&M than a red plain M&M. Whereas even the described situation is not easy to replicate numerically (by constructing another pair of such bags), a more complicated situation is to reverse chances by eating just a single red M&M from each bag. With this in mind, one can start considering several pairs of fractions compared to each other through the same relation "greater than", for example,

$$\frac{4}{9} > \frac{5}{13}, \ \frac{8}{15} > \frac{2}{9}, \ \frac{5}{11} > \frac{3}{10}, \ \frac{7}{10} > \frac{3}{5}, \ \frac{8}{12} > \frac{6}{10} \ . \tag{9.1}$$

However, subtracting the unity from each of the numeral (numerator/denominator) in all the pairs in (9.1) yields the same 'greater than' relation between the so modified fractions. Indeed,

$$\frac{3}{8} > \frac{4}{12}, \ \frac{7}{14} > \frac{1}{8}, \ \frac{4}{10} > \frac{2}{9}, \ \frac{6}{9} > \frac{2}{4}, \ \frac{7}{11} > \frac{5}{9} \ . \tag{9.2}$$

One can try many other pairs of proper fractions, most likely to the same effect. From here, one can conjecture that all pairs of fractions with numerators greater than one would possess this property; that is, one can be tempted to make the following generalization (based on empirical induction):

If $\dfrac{a}{b} > \dfrac{c}{d}$, where $1 < a < b$, $1 < c < d$ are integers, then $\dfrac{a-1}{b-1} > \dfrac{c-1}{d-1}$.

9.4.2 *From a counter-example to its conceptualization*

The above generalization, however, is not true. Here is a counter-example: whereas $4/20 > 5/26$, the subtraction of the unity from numerators and denominators yields $3/19 < 4/25$, the opposite inequality. However, to understand how this counter-example, defying the above (erroneous) conjecture, was constructed or to find another one of that kind is not easy. So, one can distinguish between counter-examples stemming from conceptual understanding of a situation and those stemming from a purely algorithmic thinking. Below, a pair of proper (non-unit) fractions $(\dfrac{a}{b}, \dfrac{c}{d})$ satisfying the inequalities $\dfrac{a}{b} > \dfrac{c}{d}$ and $\dfrac{a-1}{b-1} < \dfrac{c-1}{d-1}$ will be referred to as jumping fractions.

In order to construct a family of jumping fractions, let us start with a unit fraction, say, 1/5. We have $1/5 = 5/25 > 5/26$ and $1/5 = 4/20$. Therefore, $4/20 > 5/26$, an inequality developed constructively. However, $(4-1)/(20-1) < (5-1)/(26-1)$, that is, $3/19 < 4/25$. This algorithm of constructing a counter-example can be generalized as follows.

The unit fraction $\dfrac{1}{n}$ can be represented as either $\dfrac{m}{mn}$ or $\dfrac{n}{n^2}$, where integer $1 < m < n$. Because $\dfrac{n}{n^2} > \dfrac{n}{n^2+1}$ (the smaller the denominator, the larger the fraction), we have $\dfrac{m}{mn} > \dfrac{n}{n^2+1}$. Then the

inequality $\dfrac{m-1}{mn-1} < \dfrac{n-1}{n^2}$ holds true. Indeed, both n and m are integers

such that $n - m > 0$. Therefore, the smallest value of $n - m$ is equal to

one; that is, $n - m - 1 \geq 0$. The last inequality allows for the difference

between the fractions $\dfrac{n-1}{n^2}$ and $\dfrac{m-1}{mn-1}$ to be estimated as follows:

$$\frac{n-1}{n^2} - \frac{m-1}{mn-1} = \frac{mn^2 - n - mn + 1 - mn^2 + n^2}{n^2(mn-1)}$$

$$= \frac{n^2 - n - mn + 1}{n^2(mn-1)} = \frac{n(n-m-1)+1}{n^2(mn-1)} \geq \frac{1}{n^2(mn-1)} > 0.$$

As another example, consider the inequality 1/6 > 6/37 which is
equivalent to 5/30 > 6/37. The last two fractions are jumping fractions as
4/29 < 5/36. Note that the differences 5/30 – 6/37 = 1/222 and 5/36 –
4/29 = 1/1044 demonstrate how a pair of jumping fractions, already
being close to each other, turns into a much closer pair (Fig. 9.2).

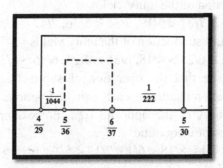

Fig. 9.2. The jumps make fractions closer to each other.

9.4.3 *A family of jumping fractions found by a teacher candidate*

This is only one family of jumping fractions that serve as a counter-
example defying the generalization about the above property of fractions
formulated at the conclusion of section 9.4.1. Yet, to find such a family
is not easy. Even continuing the subtraction of unity from the numerators
and denominators in (9.2), one cannot reach a pair in the opposite

inequality relationship. Nonetheless, one elementary teacher candidate found another system through which such pairs can be gradually developed. She started with the number 7 (her choice of 7 was not explained; after the author's analysis, a rationale for 7 is provided below) and developed fractions with the numerator 7 that are equal to consecutive unit fractions starting from 1/2, thereby writing down 7/14, 7/21, 7/28, 7/35, and so on. She then developed smaller fractions with the numerator 9 (that is, two greater than 7) and a two-digit denominator so that the corresponding fractions 9/19, 9/29, 9/39, 9/49, and so on, have been squeezed between two consecutive unit fractions as the benchmark fractions; that is,

$$\frac{1}{3} < \frac{9}{19} < \frac{1}{2}, \ \frac{1}{4} < \frac{9}{29} < \frac{1}{3}, \ \frac{1}{5} < \frac{9}{39} < \frac{1}{4}, \ \frac{1}{6} < \frac{9}{49} < \frac{1}{5},$$ and so on. Then,

taking into account that $\frac{1}{2} = \frac{7}{14}, \ \frac{1}{3} = \frac{7}{21}, \ \frac{1}{4} = \frac{7}{28}, \ \frac{1}{5} = \frac{7}{35}$, it turned out

that for the inequalities between pairs of fractions
$$9/19 < 7/14, \ 9/29 < 7/21, \ 9/39 < 7/28, \ 9/49 < 7/35$$
the first subtraction of the unity yields
$$8/18 < 6/13, \ 8/28 < 6/20, \ 8/38 < 6/27, \ 8/48 < 6/34;$$
the second subtraction of the unity yields
$$7/17 < 5/12, \ 7/27 < 5/19, \ 7/37 < 5/26, \ 7/47 < 5/33;$$
and, finally, the third subtraction of the unity yields
$$6/16 > 4/11, \ 6/26 > 4/18, \ 6/36 > 4/25, \ 6/46 > 4/32.$$

One can see that the operation of subtracting the unity from numerators and denominators in each pair of fractions leads after three steps to an inequality of the opposite sign. For example, we have the following sequence of inequalities
$$9/19 < 7/14, \ 8/18 < 6/13, \ 7/17 < 5/12, \ 6/16 > 4/11,$$
and the pair $(\frac{7}{17}, \frac{5}{12})$ represents jumping fractions. Likewise, the pairs

$(\frac{7}{27}, \frac{5}{19}), (\frac{7}{37}, \frac{5}{26})$ and $(\frac{7}{47}, \frac{5}{33})$ represent jumping fractions. That is, the pairs of fractions (7/17, 5/12), (7/27, 5/19), (7/37, 5/26), and (7/47, 5/33) are representatives of another family of pairs of fractions that serve as a counter-example to the erroneous generalization (given at the end of section 9.4.1) which was motivated by relations (9.1) and (9.2).

9.4.4 *Conceptualizing the teacher candidate's choice of seven*

Consider the following two sequences of fractions

$$\frac{7}{17}, \frac{7}{27}, \frac{7}{37}, \frac{7}{47}, \dots \text{ and } \frac{5}{12}, \frac{5}{19}, \frac{5}{26}, \frac{5}{33}, \dots .$$

Alternatively the two sequences can be written in an algebraic form

$$x_n = \frac{7}{10n+7} \text{ and } y_n = \frac{5}{7n+5}, n = 1, 2, 3, \dots . \text{ The inequality}$$

$$\frac{7}{10n+7} < \frac{5}{7n+5} \tag{9.3}$$

is equivalent to $49n + 35 < 50n + 35$ whence $n > 0$ – a true inequality.

At the same time, reducing each term of inequality (9.3) by the unity yields $\dfrac{6}{10n+6} > \dfrac{4}{7n+4}$ which is equivalent to $42n + 24 > 40n + 24$ whence $2n > 0$ – a true inequality. That is, a family of jumping fractions can be defined as follows

$$(\frac{7}{10n+7}, \frac{5}{7n+5}) , n = 1, 2, 3, \dots .$$

This algebraic definition provides a conceptualization for the teacher candidate's choice of the number 7 in defining jumping fractions.

In order to find another family of jumping fractions proceeding from inequality (9.3), consider the inequality

$$\frac{a}{10n+a} < \frac{a-2}{(n+1)a-2} \tag{9.4}$$

which is true when $a = 7$ for $n = 1, 2, 3, \dots$. Are there other values of a for which it is also true? When $n \geq 1$ and $a \geq 4$ (to allow for $a - 2 \geq 2$), inequality (9.4) is equivalent to

$$(n+1)a^2 - 2a < 10na - 20n + a^2 - 2a \text{ or } na^2 - 10na + 20n < 0$$

whence

$$a^2 - 10a + 20 < 0. \tag{9.5}$$

One can check to see that inequality (9.5) has the following integer solutions only: $a \in \{4, 5, 6, 7\}$. Now, one has to find out which one of the four values of a satisfies the inequality

$$\frac{a-1}{10n+a-1} > \frac{a-3}{(n+1)a-3}. \tag{9.6}$$

Because $a \geq 4$, it follows from (9.6) that

$(n+1)a^2 - 3a - (n+1)a + 3 > 10na - 30n + a^2 - 3a - a + 3$ or

$na^2 - 11na + 30n > 0$ whence $a^2 - 11a + 30 > 0$.

The last inequality holds true when $a < 5$ or $a > 6$. Therefore, inequalities (9.4) and (9.6) hold true simultaneously for any natural number n and $a = 4$ or $a = 7$. No wonder, the teacher candidate started with the number 7. As it turned out, her approach works for the number 4 as well. Indeed, setting $a = 4$ and $n = 1, 2, 3,$ and 4 in (9.4) and (9.6) yields, respectively,

$$4/14 < 2/6, \ 4/24 < 2/10, \ 4/34 < 2/14, \ 4/44 < 2/18,$$

and

$$3/13 > 1/5, \ 3/23 > 1/9, \ 3/33 > 1/13, \ 3/43 > 1/17.$$

One can check to see that other values of n work with $a = 4$ as well.

9.5 A Counter-Example and Empirical Induction

One of the activities recommended by New York State Department of Education [1996] to be used with first/second grade students deals with finding all possible arrangements of colors among six two-color counters, red and yellow. The goal of this activity is to use the first-order symbols (counters) as the means of transition to the second-order symbolism (de-contextualization) – determining the number of representations of six as a sum of whole numbers in all possible orders (including self-representation). As discussed in Chapter 3, section 3.5.2 (see also Chapter 5, section 5.6) there are 64 different ways two-color counters can be arranged in different alterations of colors and 32 ways of representations of six through the sums with (differently ordered) whole number addends. One way to comprehend that 64 can be found by raising the number of colors, 2, to the power equal to the number of counters, 6, is through empirical induction. To this end, one has to start with one counter for which there are two ways of having different arrangements of colors, then move to two counters for which there are four ways the counters can be arranged and then guess the number of different arrangements of three counters. For example, experimentally, one can find that for one counter there are two combinations, R and Y; for two counters there are four combinations, R&R, Y&Y, R&Y, and Y&R. From here, teacher candidates typically offer, not without doubt though, several guesses (Fig. 9.3): six combinations, nine combinations, twelve combinations, and eight combinations (this guess usually is the last one to be offered).

How can one explain that, for example, the guess "six combinations" is incorrect without actually using three counters? In other words, how can one provide a counter-example to this guess? First of all, one has to explain the rationale behind the guess. The guess "six" is supported by the relations $2 \cdot 1 = 2$ and $2 \cdot 2 = 4$ (where the first factor represents the number of sides and the second factor represents the number of counters). This yields $2 \cdot 3 = 6$ and the guess is not easy to defy. The guess "nine" appears as the square of 3 replicating the obvious $2^2 = 4$, which can be immediately rejected because $1^2 \neq 2$. The guess "twelve" is supported through multiplicative reasoning: noticing that $2 \cdot 2 = 4$ and, therefore $3 \cdot 4 = 12$; this guess is not easy to defy as well.

The guess "eight" is more complicated as it follows from raising (hidden in the table) the number of colors, 2, to the power 3 – the number of counters (this matches the relations $2^1 = 2$ and $2^2 = 4$). So, a counter-example that addresses all incorrect guesses can only be developed through a physical experiment with three two-color counters. This experiment can be based on the idea that transition from two counters to three counters can be arranged by adding first a red counter and then a yellow counter to (the above-mentioned) four combinations: R&R&R, Y&Y&R, R&Y&R, Y&R&R; and R&R&Y, Y&Y&Y, R&Y&Y, Y&R&Y. This doubling pattern continues as the number of counters increases, thereby leading to the number 64 as the sixth power of two in the case of six two-color counters.

Counters	1	2	3	...
Combs of R & Y	2 experiment	4 experiment	6, 9, 12, 8 guesses	

Fig. 9.3. From experimenting to guessing.

9.6 Counter-Example as a Tool for Conceptual Development

Another goal of using counter-examples is to help learners of mathematics articulate a certain property. For example, a third grade student was asked to show 1/4 of a square. In response to this request for information, the third grader presented a square divided into four (apparently) equal pieces (Fig. 9.4). However, the child was not able to respond to a more intelligent query: why does the piece shown represent 1/4? An answer to this query requires one to articulate the following fundamental requirement for fractional parts: there are four parts each of which has the same size. A counter-example (or non-example) of 1/4 can be helpful in explaining why the shaded part in Fig. 9.4 represents 1/4. Having in mind that, as Sugarman [1987] noted with a reference to Piaget, "children initially judge numerocity on the basis of spatial extent" (p. 138), one can draw a new sketch as shown in Fig. 9.5 and ask whether the shaded part represents 1/4. In the author's experience, the child answered "no" to this question and added: "the parts are not equal". That is, a counter-example of 1/4 helped the child to articulate conceptual meaning of this fraction in terms of the part/whole relationship. Furthermore, because already on the second month of life "the child's smile reveals that he recognizes familiar voices or faces whereas strange sounds or images astonish him" [Piaget, 1954, p. 5], one can interpret the counter-example of 1/4 provided in Fig. 9.5 as a "strange image" which for the 8-year old prompted conceptualization.

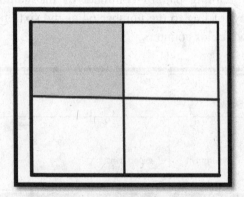

Fig. 9.4. Child's representation of 1/4.

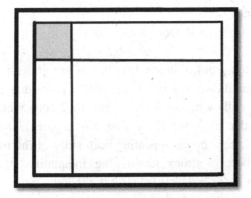

Fig. 9.5. A counter-example of 1/4.

9.7 Counter-Example in Explaining the Meaning of Negative Transfer

The third pedagogical purpose of using counter-examples is to avoid the so-called negative transfer, sometimes referred to as Einstellung effect (Chapter 7, section 7.4, Remark 7.4), in problem solving when experience with similar problems results in "habitual rigidity of thinking" [Radatz, 1979, p. 167]; that is, uncritically applying a problem-solving strategy learned in one context to a different context. In order to avoid Einstellung effect in the learning of mathematics, one needs to be able to recognize and comprehend conceptual change "when the new information to be learned comes in conflict with the learners' prior knowledge" [Vosniadou and Verschaffel, 2004, p. 445]. Often, conceptual change in mathematical problem solving is difficult to recognize because, as Pólya [1962] put it, "Human nature prompts us to repeat a procedure that has succeeded before in a similar situation" (p. 63).

In what follows, several instances of negative transfer will be considered. One instance deals with the extension of the domain of positive numbers to include also negative numbers in the context of dealing with inequalities. Another instance includes an incorrect application of proportional reasoning to the situation that is governed by additive reasoning.

9.7.1 *Inequalities*

Let $a > b > 0$ and $c > d > 0$. Does the inequality

$$ac > bd \qquad\qquad (9.7)$$

hold true? A visual proof of inequality (9.7) is presented in Fig. 9.6 in the case of the inequalities $5 > 3$ and $3 > 2$. One can see that 5 repetitions of 3 cells yield 15 cells whereas 3 repetitions of 2 cells yield only 6 cells. Formally, in order to prove (9.7), one has to consider the difference $ac - bd$ in which $ac > bc$ (as repeating both sides of the inequality $a > b$ the same number, c, times retains the inequality sign). Therefore, $ac - bd > bc - bd = b(c - d) > 0$ because $b > 0$ and $c > d$.

However, when allowing inequalities to include both positive and negative numbers, multiplying two inequalities of the same sign may or may not retain the original sign. For example, as shown in Fig. 9.7, the inequalities $3 > 2$ and $-3 > -5$ can be multiplied so that $3 \cdot (-3) > 2 \cdot (-5)$. Note that by $3 \cdot (-3)$ we mean repeating a group of three cells three times but to the left of the origin where comparison of products follows the rule 'the smaller the space the bigger the value'. Therefore, because $9 < 10$ it follows that $-9 > -10$. On the contrary, as shown in Fig. 9.8, it follows from the inequalities $3 > 1$ and $-3 > -5$ that $3 \cdot (-3) < 1 \cdot (-5)$. Likewise, the inequalities $3 > 2$ and $-3 > -4$ imply $3 \cdot (-3) < 2 \cdot (-4)$. How can this phenomenon of diverse outcomes be explained?

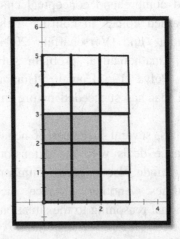

Fig. 9.6. The shaded part represents the product of 2 and 3.

Consider now the case $a > b > 0$ and $0 > c > d$; that is, the first relation compares positive numbers and the second relation compares negative numbers. Which of the two inequalities, $ac < bd$ or $ac > bd$, does hold true? As shown in Fig. 9.7 and Fig. 9.8, the outcome of multiplying the two pairs of inequalities depends on the absolute values of their components. Consider the difference $ac - bd$. We have $ac < 0$ and $-bd > 0$. The value of $ac + (-bd)$, that is, the sum of a negative number and a positive number, depends on the relation between $|ac|$ and $|bd|$, that is, between the absolute values of the numbers. In Fig. 9.7 we have $|ac| = 9$ and $|bd| = 10$. Therefore, the sum $ac + (-bd) > 0$. In Fig. 9.8, we have $|ac| = 9$ and $|bd| = 5$. Therefore, the sum $ac + (-bd) < 0$. That is, when $a > b > 0$ and $0 > c > d$ the inequalities $3 > 2$ and $-3 > -5$ serve as a counter-example to the inequality $ac < bd$ and the inequalities $3 > 1$ and $-3 > -5$ serve as a counter-example to the inequality $ac > bd$.

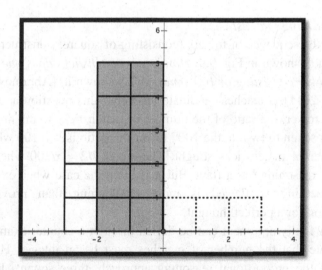

Fig. 9.7. The shaded part represents $2 \cdot (-5)$.

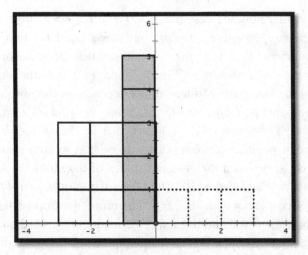

Fig. 9.8. The shaded part represents $1 \cdot (-5)$

9.7.2 *Counting matchsticks*

Consider the sequence of towers (consisting of squares constructed from matchsticks) shown in Fig. 9.9. *How many matchsticks does one need to build a tower consisting of 100 squares?* As shown in [Abramovich and Brouwer, 2011], a teacher candidate answered this question as follows: in the 2nd tower, the ratio of the number of matchsticks to the number of squares is seven to two; in the 100th tower this ratio is x to 100 where x is the number of matchsticks sought. Therefore, $7/2 = x/100$ whence $x = 350$. This reasoning has a flaw. But it is often the case when explaining why something is flawed is more challenging than providing a justification for a perfect thought.

A faulty argument can be defied through a counter-example. To begin, note that the number of matches must be an integer. However, following the proportional reasoning approach, three squares have ten matches and the proportion $10/3 = x/100$ yields $x = 1000/3$ – not only a different answer but not even an integer! Such recourse to a counter-example does indicate that the proportional reasoning approach is not correct, but does not explain why it is incorrect. This explanation may be based on the analysis of the relationship between the matchsticks and squares. There is no evidence that the number of matchsticks and the number of squares are in the same ratio (Fig. 9.9). This is clear already by comparing the first two towers in the sequence: one square requires four matchsticks, and two squares require seven matchsticks. It is because of the bottom square, which requires four matchsticks, that proportional reasoning approach has a flaw. If the first square were built out of three matches, then proportional reasoning would work fine (Fig. 9.10). Indeed, each new square requires three matches, so that 100 squares would require 300 matchsticks having the ratio of matchsticks to squares equal 3 to 1.

By the way, the number of matchsticks in the 100-square towers in Fig. 9.9 and Fig. 9.10 differ by one and this observation may immediately bring about 301 as the right answer to the original question. Note that if proportional reasoning were applied to the (full) 50-square tower built out of 151 ($= 3 \cdot 50 + 1$) matchsticks, the proportion $151/50 = x/100$ yields $x = 302$ – still an incorrect answer but it is very close to the right one, 301. The use of proportional reasoning approach might also serve as a counter-example to a possible conclusion that such reasoning always results in a non-integer answer. In fact, the 50-square tower is the largest one that can serve as such a counter-example because 50 is the largest proper divisor of 100, that is, the largest number different from 100 that divides 100.

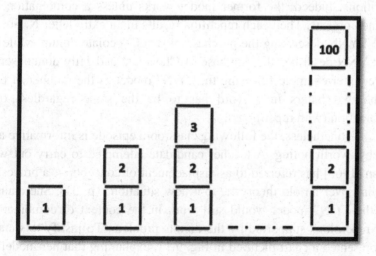

Fig. 9.9. Matches and squares are not in the same ratio.

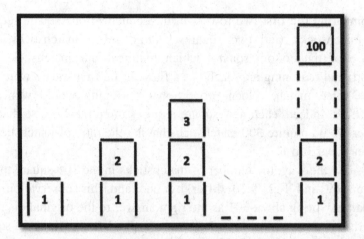

Fig. 9.10. Matches and squares are in the same ratio.

9.8 Transition from Combinations without Repetitions to Combinations with Repetitions

As was shown in Chapter 2 (section 6), extending the strategy of counting combinations as permutations in a word using the "Y/N" model to that of counting combinations with repetitions using the "Y/B" model provided the uniformity of the description of combinations with repetitions. Indeed, the former model works unless a combination has repeated objects. Then each repetition results in an extra letter N. So, the word YYNN describing the purchase of two Chocolate donuts while the word YNN describes the purchase of Chocolate and Jelly donuts serves as a counter-example for using the "Y/N" model as the number of each of the two letters in a word has to be the same regardless of a combination (with repetitions).

Nonetheless, the following classroom episode is informative and, thereby, worth noting. A teacher candidate attempted to carry out what Schon [1963] has referred to as displacement of concepts – "a process of carrying over an old theory ... to a new situation" (p. 31). She claimed that the "Y/N" model would still work in the context of combinations with repetitions supported by the donuts problem. To justify this claim, she presented a chart pictured in Fig. 9.11, explaining that her model not only includes three types of donuts, but, better still, it allows for a six letter word with two Y's and four N's to describe any selection of two donuts. That is, in her chart (Fig. 9.11) each type of donuts is listed twice (because two donuts have to be selected) so that the words YNYNNN and YNNYNN in the first and the second rows, respectively, represent permutations of letters in the same six-letter word representing, respectively, the selection of Chocolate and Jelly donuts in the former case and two Chocolate donuts in the latter case.

Once again, it is easier to explain why something is correct rather than incorrect. One way to provide an explanation in the latter case is through a counter-example. The presence of a physical model shown in Fig. 2.10 of Chapter 2 is helpful in this regard, as the new symbolic model (Fig. 9.11) suggested by the teacher candidate can be verified experimentally. Indeed, the number of permutations of letters in the word YYNNNN is equal to $\dfrac{6!}{2! \, 4!} = 15$. Yet, Fig. 2.10 of Chapter 2 indicates that there exist only six ways to select two donuts out of three types. In this case, a physical model serves as a counter-example to a theoretical model proposed by a teacher candidate. This episode points at the importance of a physical experiment as a means of verifying the correctness of a theoretical model.

CH	GL	JEL	CH	GL	JEL
Y	N	Y	N	N	N
Y	N	N	Y	N	N
N	N	Y	N	Y	N
Y	Y	N	N	N	N
N	N	Y	Y	N	N
N	N	N	N	Y	Y

Fig. 9.11. A theoretical model contradicts a true experiment.

9.9 Missing Fibonacci Numbers

Consider the following activity designed to introduce Fibonacci[31] numbers to elementary teacher candidates.

Determine the number of different arrangements of one, two, three, four, and so on two-color (red/yellow) counters in which no two red counters appear back to back.

By using the counters (Fig. 9.12), a teacher candidate arranged one counter in two ways (R, Y), two counters – in three ways (RY, YR, YY), three counters – in five ways (RYR, RYY, YRY, YYR, YYY). Using the W^4S principle (Chapter 4), she developed a chart shown in Fig. 9.13 in which the first three numbers in the second row (2, 3, and 5) reflected data shown in Fig. 9.12, whereas the rest of the numbers in that row (8, 12, 17, 23, and 30) were found by inductive reasoning. She explained the appearance of the guessed numbers as follows: the sum of the number of counters, n, and the number of their arrangements, $A(n)$, is equal to the next number $A(n + 1)$ in the second row (indeed, the experimental numbers follow this observation: $3 = 1 + 2$ and $5 = 2 + 3$). Thus, other numbers were obtained:

$8 = 3 + 5$, $12 = 4 + 8$, $17 = 5 + 12$, $23 = 6 + 17$, and $30 = 7 + 23$.

When asked to explain the meaning of the observed pattern within experimental data, she was unsure. This was not surprising, as the pattern was confirmed by very limited experiential evidence and, consequently, led to an incorrect generalization. Indeed, using the counters, one can show that, although the number 8 does show the correct number of the corresponding arrangements (Fig. 9.14), already the number 12 does not. It is interesting to note that the sequence 2, 3, 5, 8, 12, 17, 23, 30, ... can be connected to triangular numbers if each of the terms is diminished by two: 0, 1, 3, 6, 10, 15, 21, 28, ... This is also not surprising because it is how triangular numbers are defined. So, a counter-example is of the critical importance here. Put another way, a counter-example can convince one that a pattern found does break down. A similar example

---

[31] Leonardo Fibonacci (1170–1250) – an Italian mathematician credited with the introduction of Hindu-Arabic number system into the Western world.

deals with the largest number of regions formed by the intersecting lines connecting points on a circle [Maier, 1988]: a line connecting two points can form two regions, lines connecting three points can form four regions, lines connecting four points can form eight regions, lines connecting five points can form 16 regions. From here, by empirical induction, it is quite plausible to conclude that the lines connecting six points on a circle form 32 regions. While this is a good guess, it is not correct: the number of such regions is 31.

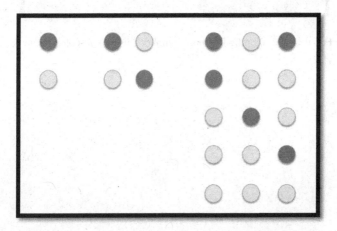

Fig. 9.12. Developing Fibonacci numbers 2, 3, and 5 using two-color counters.

counters	1	2	3	4	5	6	7	8
arrangements	2	3	5	8	12	17	23	30

Fig. 9.13. Adding numbers from different rows yields an error on the 4th step.

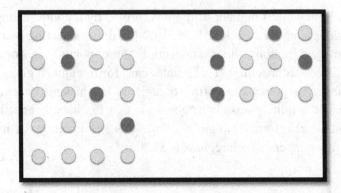

Fig. 9.14. Developing Fibonacci number 8 using two-color counters.

Bibliography

Abramovich, S. (2010a) *Topics in Mathematics for Elementary Teachers: A Technology-Enhanced Experiential Approach*, (Information Age Publishing, Charlotte, NC).

Abramovich, S. (2010b). Modeling as isomorphism: the case of teacher education. In R. Lesh, P. L. Galbraith, C. R. Haines and A. Hurford (Eds), *Modeling Students' Mathematical Modeling Competencies: ICTMA 13*, (Springer, New York) pp. 501–510.

Abramovich, S. (2012). Counting and reasoning with manipulative materials: A North American perspective. In N. Petrovic (Ed.), *The Interfaces of Subjects Taught in the Primary Schools and on Possible Models of Integrating Them*, (The University of Novi Sad Faculty of Education Press, Sombor, Serbia) pp. 9–20.

Abramovich, S. (2014). *Computational Experiment Approach to Advanced Secondary Mathematics Curriculum*, (Springer, Dordrecht, The Netherlands).

Abramovich, S. (2015). Mathematical problem posing as a link between algorithmic thinking and conceptual knowledge, *The Teaching of Mathematics*, 18(2), pp. 45–60.

Abramovich, S. (2016). Einstellung effect and the modern day mathematical problem solving and posing. In K. Newton (Ed.), *Problem Solving: Strategies, Challenges and Outcomes*, (Nova Science Publishers, New York) pp. 51–64.

Abramovich, S. and Brouwer, S. (2006). Hidden mathematics curriculum: a positive learning framework, *For the Learning of Mathematics*, 26(1), pp. 12–16, 25.

Abramovich, S. and Brouwer, P. (2011). Where is the mistake? The matchstick problem revisited, *PRIMUS*, 21(1), pp. 11–25.

Abramovich, S. and Brown, G. (1999). From measuring to formal demonstration using interactive computational geoboards and recurrent electronic charts, *Journal of Computers in Mathematics and Science Teaching*, 18(2), pp. 105–134.

Abramovich, S., Easton, J. and Hayes, V. O. (2012). Parallel structures of computer-assisted signature pedagogy: the case of integrated spreadsheets, *Computers in the Schools*, 29(1–2), pp. 174–190.

Abramovich, S. and Leonov, G. A. (2009). Spreadsheets and the discovery of new knowledge, *Spreadsheets in Education*, 3(2), Article 1. Available at: http://epublications.bond.edu.au/esie/vol3/iss2/1.

Abramovich, S. and Sugden, S. (2008). Diophantine equations as a context for technology-enhanced training in conjecturing and proving, *PRIMUS*, 18(3), pp. 257–275.

Abrams, A. D. and Paris, M. T. (1992). The probability that $(a, b) = 1$, *The College Mathematics Journal*, 23(1), p. 47.

Adler, J. and Venkat, H. (2014). Mathematical knowledge for teaching. In S. Lerman (Ed.), *Encyclopedia of Mathematics Education*, (Springer, Dordrecht, The Netherlands) pp. 385–388.

Advisory Committee on Mathematics Education. (2007). *Mathematical Needs of 14–19 Pathways*, (The Royal Society, London).

Advisory Committee on Mathematics Education. (2011). *Mathematical Needs: The Mathematical Needs of Learners*, (The Royal Society, London). Available at http://www.nuffieldfoundation.org/sites/default/files/files/ACME_Theme_B_final.pdf.

Ahlgren, S. and Ono, K. (2001). Addition and counting: the arithmetic of partitions, *Notices of the American Mathematical Society*, 48(9), pp. 978–984.

Akinwunmi, K., Höveler, K. and Schnell, S. (2014). On the importance of subject matter in mathematics education: A conversation with Erich Christian Wittmann, *Eurasia Journal of Mathematics, Science, and Technology Education*, 10(4), pp. 357–363.

Aleksandrov, A. D. (1963). A general view of mathematics. In A. D. Aleksandrov, A. N. Kolmogorov, and M. A. Lavrent'ev (Eds), *Mathematics: Its Content, Methods and Meaning*, (MIT Press, Cambridge, MA) pp. 1–64.

Ahlfors, L. V. (1962). On the mathematics curriculum of the high school [Memorandum], *American Mathematical Monthly*, 69(3), pp. 189–193.

Arbuthnot, J. (1710). An argument for divine providence taken from the constant regularity observed in the births of both sexes,

Philosophical Transactions of the Royal Society of London, 27 (pp. 186-190).

Archimedes. (1912). *The Method of Archimedes*, T. L. Heath (Ed.), (Cambridge University Press, Cambridge, England).

Arnold, V. I. (2005). *Experimental Mathematics*, (Fazis, Moscow), In Russian.

Arnold, V. I. (2015). *Lectures and Problems: A Gift to Young Mathematicians*, (American Mathematical Society, Providence, RI).

Association of Teachers of Mathematics. (1967). Notes on Mathematics in Primary Schools, (Cambridge University Press, London).

Australian Association of Mathematics Teachers. (2006). *Standards for Excellence in Teaching Mathematics in Australian Schools* [On-line materials]. Available at:www. aamt. edu.au.

Balacheff, N. (1988). Aspects of proof in pupils' practice of school mathematics, In D. Pimm (Ed.), *Mathematics, Teachers, and Children* (Hodder and Stoughtonm, London) pp. 216—238

Ball, D. L. (1992). Magical hopes: manipulatives and the reform of mathematics education, *American Educator*, 16(2), pp. 14–18, 46–47.

Baumert, J., Kunter, M., Blum, W., Brunner, M., Voss, T., Jordan, A., Klusmann, U., Krauss, S., Neubrand, M. and Tsai, Y.-M. (2010). Teachers' mathematical knowledge, cognitive activation in the classroom, and student progress, *American Educational Research Journal*, 47(1), pp. 133–180.

Bennett, D. J. (1998). *Randomness*, (Harvard University Press, Cambridge, MA).

Boas R. P. (1971). Calculus as an experimental science, *American Mathematical Monthly*, 78(6), pp. 664–667.

Boggan, M., Harper, S. and Whitmire, A. (2010). Using manipulatives to teach elementary mathematics, *Journal of Instructional Pedagogies*, 3(1), pp. 1–10.

Borwein, J. and Bailey, D. (2004). *Mathematics by Experiment: Plausible Reasoning in the 21st Century*, (A K Peters, Natick, MA).

Boyer, C. (1968). *A History of Mathematics*, (John Wiley & Sons, New York).

Brodinsky, B. (1977). Back to the basics: the movement and its meaning, *Phi Delta Kappan*, 58(7), pp. 522–527.

Carreira, S., Jones, K., Amado, N., Jacinto, H. and Nobre, S. (2016). *Youngsters Solving Mathematical Problems with Technology*, (Springer, Dordrecht, The Netherlands).

Chazan, D. (1993). High school geometry students' justification for their views of empirical evidence and mathematical proof, *Educational Studies in Mathematics*, 24(4), pp. 359–387.

Christou, C., Mousoulides, N., Pittalis, M. and Pitta-Pantazi, D. (2004). Proofs through exploration in dynamic geometry environments, *International Journal of Science and Mathematics Education*, 2(3), pp. 339–352.

Clements, D. H. (1999). 'Concrete' manipulatives, concrete ideas, *Contemporary Issues in Early Childhood Education*, 1(1), pp. 45–60.

Common Core State Standards. (2010). *Common Core Standards Initiative: Preparing America's Students for College and Career* [On-line materials]. Available at: http://www.corestandards.org.

Conference Board of the Mathematical Sciences. (2001). *The Mathematical Education of Teachers*, (The Mathematical Association of America, Washington, DC).

Conference Board of the Mathematical Sciences. (2012). *The Mathematical Education of Teachers II*, (The Mathematical Association of America, Washington, DC).

Cooke, R. (2010). Life on the mathematical frontier: legendary figures and their adventures, *Notices of the American Mathematical Society*, 57(4), pp. 464–475.

Cuoco, A. (2005). *Mathematical Connections: A Companion for Teachers and Others*, (The Mathematical Association of America, Washington, DC).

David, F. N. (1970). Dicing and gaming (a note on the history of probability). In E. S. Pearson and M. G. Kendall (Eds), *Studies in the History of Statistics and Probability*, (Griffin, London) pp. 1–17.

De Mairan, J. J. (1728). Sur le jeu de pair ou non. *Hist. de L'Acad. R. Sci. Paris*, pp. 53–57.

Department for Education (2013). National Curriculum in England: Mathematics Programmes of Study, Crown copyright. Available at: https://www.gov.uk/government/publications/national-curriculum-in-england-mathematics-programmes-of-study.

Descartes, R. (1965). *A Discourse on Method*, (J. M. Dent and Sons, London).

DeSoto, C. B., London, M. and Handel, S. (1965). Social reasoning and spatial paralogic, *Journal of Personality and Social Psychology*, 2(4), pp. 513–521.

DeTemple, D. and Robertson, J. M. (1974). The equivalence of Euler's and Pick's theorems, *Mathematics Teacher*, 67(3), pp. 222–226.

De Villiers, M. (2004). The role and function of quasi-empirical methods in mathematics, *Canadian Journal of Mathematics, Science, and Technology Education*, 4(3), pp. 397–418.

Dewey, J. (1933). *How We Think: A Restatement of the Relation of Reflective Thinking to the Education Process*, (Heath, Boston).

Dunham, W. (1999). *Euler. The Master of Us All*, (The Mathematical Association of America, Washington, DC).

Dunker, K. (1945). On problem solving, *Psychological Monographs*, 58 (whole No. 270).

Ellis, W. D. (Ed.). (1938). *A Source Book of Gestalt Psychology*, (Harcourt Brace, New York).

Felmer, P., Lewin, R., Martínez, S., Reyes, C., Varas, L., Chandía, E., Dartnell, P., López, A., Martínez, C., Mena, A., Ortíz, A., Schwarze, G. and Zanocco, P. (2014). *Primary Mathematics Standards for Pre-Service Teachers in Chile*, (Singapore, World Scientific).

Freudenthal, H. (1978). *Weeding and Sowing*, (Kluwer, Dordrecht, The Netherlands).

Freudenthal, H. (1983). *Didactical Phenomenology of Mathematical Structures*, (Reidel, Dordrecht, The Netherlands).

Gardner, J. H. (1991). "How fast does the wind travel?" History in the primary mathematics classroom. *For the Learning of Mathematics*, 11(2), pp. 17–20.

Gattegno C. (1963). *Modern Mathematics with Numbers in Colour*, (Lamport Gilbert, Reading, England).

Gattegno, C. (1971). *Geoboard Geometry*, (Educational Solutions Worldwide, New York).

Gibson, J. J. (1977). The theory of affordances. In R. Shaw and J. Bransford (Eds), *Perceiving, Acting and Knowing: Toward an Ecological Psychology*, (Lawrence Erlbaum, Hillsdale, NJ) pp. 67–82.

Goldenberg, E. P. and Cuoco, A. A. (1998). What is dynamic geometry? In R. Lehrer and D. Chazan (Eds), *Designing Learning Environments for Developing Understanding of Geometry and Space*, (Lawrence Erlbaum, Mahwah, NJ) pp. 351–367.

Guin, D. and Trouche, L. (2002). Mastering by the teacher of the instrumental genesis in CAS environments: necessity of instrumental orchestrations. *Zentralblatt für Didaktik der Mathematik (ZDM)*, 34(5), pp. 204–211.

Hanna, G. (1995). Challenges to the importance of proof, *For the Learning of Mathematics*, 15(3), pp. 42–49.

Haroutunian-Gordon, S. and Tartakoff, D. S. (1996). On the learning of mathematics through conversation, *For the Learning of Mathematics*, 16(2), pp. 2–10.

Hersh, R. (1993). Proving is convincing and explaining, *Educational Studies in Mathematics,* 24(4), pp. 389—399.

Hollebrands, K. F. (2007). The role of a dynamic software program for geometry in the strategies high school mathematics students employ, *Journal for Research in Mathematics Education*, 38(2), pp. 164–192.

Hohenwarter, M. (2002). *GeoGebra* [Computer program]. Available at: http://www.GeoGebra.org/.

Hombas, V. (2013). What's the probability of a rational ratio being irreducible?, *International Journal of Mathematical Education in Science and Technology*, 44(3), 408–410.

Hoyles, C., Healy, L. and Noss, R. (1995). Can dynamic geometry constructions replace proof or contribute to it? In C. Mammana (Ed.), *ICMI Study: Perspectives on the Teaching of Geometry for the 21st Century*, (University of Catania, Catania, Italy) pp. 101–104.

Hwang, H. and Han, H. (2013). Current national mathematics curriculum. In J. Kim, I. Han, M. Park and J. K. Lee (Eds), *Mathematics Education in Korea*, (World Scientific, Singapore) pp. 21–42.

Isaacs, N. (1930). Children's why questions. In S. Isaacs, *Intellectual Growth in Young Children*, (Routledge & Kegan Paul, London) pp. 291–349.

Jackiw, N. (1991). *The Geometer's Sketchpad* [Computer program], (Key Curriculum Press, Berkeley, CA).

Kakihana, K., Shimizu, K., Nohda, N., Hibasibara, Y. and Nakayma, K. (1994). The roles of measurement in proof problems, In J. P. da Ponte and J. F. Matos (Eds), *Proceedings of the Eighteenth International Conference for the Psychology of Mathematics Education* (volume 3), (University of Lisbon, Lisbon, Portugal) pp. 82—88.

Kadijevich, Dj. and Haapsalo, L. (2000). Two types of mathematical knowledge and their relation, *Journal für Mathematik-Didaktik (JMD)*, 21(2), pp. 139–157.

Kaufman, E. L., Lord, M. W., Reese, T. W. and Volkmann, J. (1949). The discrimination of visual number, *The American Journal of Psychology*, 62(4), pp. 498–525.

Kieran, C. and Drijvers, P. (2006). The co-emergence of machine techniques, paper-and-pencil techniques, and theoretical reflection: a study of CAS use in secondary school algebra, *International Journal of Computers for Mathematical Learning*, 11(2), pp. 205–263.

Klimchuk, S. (2010). *Counterexamples in Calculus*, (The Mathematical Association of America, Washington, DC).

Knauff, M. and May, E. (2006). Mental imagery, reasoning, and blindness, *The Quarterly Journal of Experimental Psychology*, 59(1), pp. 161–177.

Laborde, J-M. (1991). *CABRI Geometry* [Computer program], (Brooks-Cole Publishing Co, New York).

Laplace, P. S. (1774). Mémoire sur les suites récurro-récurrentes et sur leurs usages dans la théorie des hasards, *Mém. Acad R. Sci. Paris* (Savants étrangers), 6, pp. 353–371.

Luchins, A. S. (1942). Mechanization in problem solving: the effect of Einstellung. *Psychological Monographs*, 54(6, whole No. 248), (The American Psychological Association, Evanston, IL).

Luchins, A. S. and Luchins, E. H. (1970a). *Wertheimer's seminars revisited: problem solving and thinking*, volume I, (Faculty-Student Association, SUNY at Albany, Albany, NY).

Luchins, A. S. and Luchins, E. H. (1970b). *Wertheimer's seminars revisited: problem solving and thinking*, volume II, (Faculty-Student Association, SUNY at Albany, Albany, NY).

Luchins, A. S. and Luchins, E. H. (1994). The water jar experiments and Einstellung effect. Part II: Gestalt psychology and past experience. *Gestalt Theory: Official Journal of the Society for Gestalt Theory and its Applications (GTA)*, 16(4), pp. 205–259.

Lingefjärd, T. (2012). Mathematics teaching and learning in a technology rich world, In S. Abramovich (Ed.), *Computers in Education, Volume 2*, (Nova Science Publishers, New York) pp. 171–191.

Lotman, Y. M. (1988). Text within a text, *Soviet Psychology*, 24(3), pp. 32–51.

Maher, C. A. and Martino, A. M. (1996). The development of the idea of mathematical proof: a 5-year case study, *Journal for Research in Mathematics Education*, 27(2), pp. 194–214.

Maier, E. (1988). Counting pizza pieces and other combinatorial problems, *The Mathematics Teacher*, 81(1), pp. 22–26.

Marrades, R. and Gutiérrez, Á. (2000). Proofs produced by secondary school students learning geometry in a dynamic computer environment, *Educational Studies in Mathematics*, 44(1), pp. 87–125.

Mason, J. (2000). Asking mathematical questions mathematically, *International Journal of Mathematical Education in Science and Technology*, 31(1), pp. 97–111.

Mason, J. and Klimchuk, S. (2009). *Using Counter-Examples in Calculus*, (Imperial College Press, London).

Ministry of Education, Singapore. (2012). *Mathematics Syllabus, Primary One to Four*, Curriculum Planning and Development Division: Author. Available at: https://www.moe.gov.sg/docs/default-source/document/education/syllabuses/sciences/files/mathematics-syllabus-(primary-1-to-4).pdf.

Mises, R., v. (1957). *Probability, Statistics and Truth*, (Macmillan, New York).

National Curriculum Board. (2008). *National Mathematics Curriculum: Framing Paper*, (Author, Australia). Available at: http://www.acara.edu.au/verve/_resources/National_Mathematics_Curriculum_-_Framing_Paper.pdf.

National Council of Teachers of Mathematics. (1991). *Professional Standards for Teaching Mathematics*, (Author, Reston, VA).

National Council of Teachers of Mathematics. (2000). *Principles and Standards for School Mathematics*, (Author, Reston, VA).

New York State Education Department. (1996). *Learning Standards for Mathematics, Science, and Technology*, (Author, Albany, NY).

New York State Education Department. (1998). *Mathematics Resource Guide with Core Curriculum*, (Author, Albany, NY).

Nickerson, R. S. (2004). *Cognition and Chance: The Psychology of Probabilistic Reasoning*, (Lawrence Erlbaum, Mahwah, NJ).

Ontario Ministry of Education. (2005). *The Ontario Curriculum, Grades 1–8, Mathematics (revised)* [on-line materials]. Available at: http://www.edu.gov.on.ca.

Ore, Ø. (1960). Pascal and the invention of probability theory, *The American Mathematical Monthly*, 67(5), pp. 409–419.

Peled, I. and Zaslavsky. O. (1997). Counter-examples that (only) prove and counter-examples that (also) explain, *Focus on Learning Problems in Mathematics*, 19(3), pp. 49–61.

Piaget, J. (1954). *The Construction of Reality in the Child*, (Basic Books, New York).

Pólya, G. (1954). *Mathematics and Plausible Reasoning* (volume 1: *Induction and Analogy in Mathematics*), (Princeton University Press, Princeton, NJ).

Pólya, G. (1962). *Mathematical Discovery: On Understanding, Learning, and Teaching Problem Solving* (volume 1), (John Wiley & Sons, New York).

Power, D. J. (2000), A brief history of spreadsheets, *DSSResources.COM*, Available at: http://dssresources.com/history/sshistory.html.

Presmeg, N. (2006). Research on visualization in learning and teaching mathematics, In A. Gutiérrez and P. Boero (Eds), *Handbook of Research on the Psychology of Mathematics Education: Past, Present and Future*, (Sense Publishers, Rotterdam, The Netherlands) pp. 205–235.

Radatz, H. (1979). Error analysis in mathematics education, *Journal for Research in Mathematics Education*, 10(3), pp. 163–172.

Reichenbach, H. (1949). *The Theory of Probability*, (University of California Press, Berkeley, CA).

Roegel, D. (2013). A reconstruction of Joncourt's table of triangular numbers (1762). *Technical Report*. Nancy, France: Lorraine Laboratory of IT Research and its Applications. (A reconstruction of: Élie de Joncourt. De natura et præclaro usu simplicissimæ speciei numerorum trigonalium. The Hague: Husson, 1762). Available at: http://locomat.loria.fr.

Rohlin, V. A. (2013). *A Lecture about Teaching Mathematics to Non-Mathematicians. Part I*. Available at: http://mathfoolery.wordpress.com/2011/01/01/a-lecture-about-teaching-mathematics-to-non-mathematicians/.

Rudman, P. S. (2007). *How Mathematics Happened*, (Prometheus Books, Amherst, NY).

Schilling, M. F. (2012). The surprising predictability, *Mathematics Magazine*, 85(2), pp. 141–149.

Schmittau, J. and Vagliardo, J. J. (2006). Using concept mapping in the development of the concept of positional system. In A. J. Cañas and J. D. Novak (Eds), *Concept Maps: Theory, Methodology, Technology*, Proceedings of the Second International Conference on Concept Mapping, (Universidad de Costa Rica, San José, Costa Rica) pp. 590–597.

Schon, D. A. (1963). *Invention and the Evolution of Ideas*, (Social science paperbacks, London).

Shulman, L. S. (1986). Those who understand: knowledge growth in teaching, *Educational Researcher*, 15(2), pp. 4–14.

Shulman, L. S. (2005). Signature pedagogies in the professions, *Daedalus*, 134(3), pp. 52–59.

Skiena, S. (1990). *Implementing Discrete Mathematics*, (Addison-Wesley, Redwood City, CA).

Smith, D. E. (1953). *History of Mathematics, Volume II (special topics of elementary mathematics)*, (Dover, New York).

Stanton, D. and White, D. (1986). *Constructive Combinatorics*, (Springer-Verlag, New York).

Steen, L. A. and Seebach, J. A., Jr. (1995). *Counterexamples in Topology*, (Dover, New York).

Stewart, I. (1990). Change. In L. A. Steen (Ed.), *On the Shoulders of Giants*, (National Academy Press: Washington, DC) pp. 183-217.

Sugarman, S. (1987). *Piaget's Construction of the Child's Reality*, (Cambridge University Press, Cambridge, England).

Sutherland, R. (1994). The role of programming: towards experimental mathematics. In R. Biehler, R. W. Scholz, R. Sträber, and B. Winkelmann (Eds), *Didactics of Mathematics as a Scientific Discipline* (Kluwer, Dordrecht, The Netherlands) pp. 177–187.

Székely, G. J. (1986). *Paradoxes in Probability Theory and Mathematical Statistics* (Reidel, Dordrecht, Holland).

Takahashi, A., T. Watanabe, Yoshida, M., and McDougal, T. (2004). *Elementary School Teaching Guide for the Japanese Course of Study: Arithmetic (Grade 1-6)*, (Global Education Resources, Madison, NJ).

Tall, D. (1991). Intuition and rigor: the role of visualization in the calculus, In W. Zimmermann and S. Cunningham (Eds), *Visualization in Teaching and Learning Mathematics*, (The Mathematical Association of America, Washington, DC) pp. 105–119.

Tharp, R. and R. Gallimore. (1988). *Rousing Minds to Life: Teaching, Learning, and Schooling in Social Context*, (Cambridge University Press, Cambridge, England).

Tijms, H. (2012). *Understanding Probability*, (Cambridge University Press, Cambridge, UK).

Tirosh, D. and'Graeber, A. O. (1989). Preservice elementary teachers' explicit beliefs about multiplication and division, *Educational Studies in Mathematics*, 20(1), pp. 79–96.

Todhunter, I. (1949). *A History of the Mathematical Theory of Probability*, (Chelsea, New York).

Uttal, D. H., Scudder, K. V. and DeLoache, J. S. (1997). Manipulatives as symbols: a new perspective on the use of concrete objects to teach mathematics, *Journal of Applied Developmental Psychology*, 18(1), pp. 37–54.

Van De Walle, J. A. (2001). *Elementary and Middle School Mathematics: Teaching Developmentally* (4th edition), (Addison Wesley, New York).

Van De Walle, J. A., Karp, K. S. and Bay-Williams, J. M. (2010). *Elementary and Middle School Mathematics: Teaching Developmentally*, (Allyn & Bacon, Boston).

Van der Waerden, B. L. (1961). *Science Awakening*, (Oxford University Press, New York).

Vilenkin, N. Y. (1971). *Combinatorics*, (Academic Press, New York).

Vosniadou, S. and Verschaffel, L. (2004). Extending the conceptual change approach to mathematics learning and teaching, *Learning and Instruction*, 14(5), pp. 445–451.

Vos Savant, M. (1996). *The Power of Logical Thinking*, (St. Martin's Press, New York).

Vygotsky, L. S. (1962). *Thought and Language*, (MIT Press, Cambridge, MA).

Vygotsky, L. S. (1978). *Mind in Society*, (Harvard University Press, Cambridge, MA).

Vygotsky, L. S. (1987). Thinking and speech, In R. W. Rieber and A. S. Carton (Eds), *The Collected Works of L. S. Vygotsky*, vol. 1 (Plenum Press, New York) pp. 39–285.

Wagenaar, W. A. (1972). Generation of random sequences by human subjects: a critical review of literature, *Psychological Bulletin*, 77(1), pp. 65–72.

Watson, J. and Fitzallen, N. (2016). Statistical software and mathematics education: affordances for learning, In L. D. English and D. Kirshner (Eds), *Handbook of International Research in Mathematics Education* (3rd edition), (Routledge, New York), pp. 563–594.

Wertheimer, M. (1938). Gestalt theory, In W. D. Ellis (Ed.), *A Source Book of Gestalt Psychology*, (Kegan Paul, London) pp. 1–11.

Wertheimer, M. (1959). *Productive Thinking*, (Harper & Row, New York).

Wertsch, J. V. (1991). *Voices of the Mind: A Sociocultural Approach to Mediated Action*, (Harvard University Press, Cambridge, MA).

Wiles, A. (1995). Modular elliptic curves and Fermat's Last Theorem, *Annals of Mathematics*, 141(3), pp. 443–551.

Yerushalmy, M. (1990). Using empirical information in geometry: student's and designer's expectations, *Journal of Computers in Mathematics and Science Teaching*, 9(3), pp. 23–38.

Zazkis, R. and Chernoff, E. (2008). What makes a counterexample exemplary?, *Educational Studies in Mathematics*, 68(3), pp. 195–208.

Zimmermann,W. and Cunningham, S. (Eds). (1991). *Visualization in Teaching and Learning Mathematics*, (The Mathematical Association of America, Washington, DC).

Index

abstraction, 89, 112, 114
addition table, 204
additive partition, 106
algorithm, 10, 41, 42, 87
Arbuthnot, 207
Archimedes, 164, 196
area, 7, 58, 63, 229, 230
area model, 78, 88, 89, 90, 206
arithmetic mean, 68
array, 17, 117, 121, 165
assisted performance, 130
automatism, 14, 130
average, 146

benchmark fractions, 80, 84, 86
Bernoulli, 208
Bernoulli trials, 208, 211
Bertrand's Box Paradox, 214
binomial coefficients, 121
boundary conditions, 119, 178, 179

CABRI Geometry, 163
Cardano, 198
change of unit, 85, 95, 97
coin, 208, 214, 220
collateral learning, 196
combination, 24, 54
combinations with repetitions, 27,
     28, 246
combinatorics, 18, 22, 231
common denominator, 83
common sense, 9, 33, 34, 106
comprehension, 52
concept map, 128
conceptual development, 37

conceptual understanding, 10, 11,
     13, 14, 16, 24, 57, 73, 75, 80,
     147, 202
conjecture, 163, 164
connections, 12, 15, 157
contextualization, 87
counter-example, 37, 39, 227
counting, 17, 60, 71
creativity, 130, 177

De Mairan, 212
De Méré, 209, 219, 223
decimal notation, 100
decimal representation, 99
decimal system, 65
de-contextualization, 87, 227
deductive reasoning, 28, 34, 164,
     193
Descartes, 198
Dewey, 131
die, 206, 209, 220
Dirichlet, 225
displacement of concepts, 29, 246
distributive property, 88
diversity of methods, v, 18, 27, 30,
     44, 80, 114, 142, 174
division, 11, 82, 89

early algebra, 50
Egyptian papyrus roll, 166
Einstellung effect, 177, 178, 179,
     227, 241
Élie de Joncourt, 72
empirical induction, 126, 164, 173,
     234, 238, 249

engagement, 51
equally likely outcomes, 203, 205
equilateral triangle, 46
Eratosthenes, 164
estimation, 84
Euclidean algorithm, 11
Euler, 8, 105, 187, 225
event, 199, 203, 224
experiment, 200, 203
experimental mathematics, 35
experimental probability, 75, 221

Fermat, 18
Fermat's Last Theorem, 132
Ferrers-Young diagrams, 117
Fibonacci, 248
Fibonacci numbers, 11
first-order symbols, 35, 49, 50, 51,
    63, 66, 73, 74, 98, 140, 151,
    182, 199, 232, 238
fixed mind-set, 14
formation, 51
fraction, 82, 85, 101, 171, 205, 224
fraction circles, 75, 78

Galileo, 217
games of chance, 18, 209, 220
Gauss, 225
geoboard, 7, 165, 171, 229
GeoGebra, 163
geometric series, 99
geometry, 163
Gestalt psychology, 28, 136, 170
gnomon, 71
Goldbach's Conjecture, 132
greatest common divisor, 171, 224
greatest integer function, 44

growth mind-set, 14
guess, 239, 249

hidden mathematics curriculum,
    129, 144
historical connections, 63

inequality, 78, 80, 233, 235, 237
insight, 6, 14, 29, 74
intuition, 27, 33
intuitive understanding, 34
invert and multiply rule, 95
irreducible fraction, 224
isomorphic model, 69
isosceles triangle, 46

jumping fractions, 234

kindergarten classroom, 42

Laplace, 212
Law of Large Numbers, 208, 221
long division, 9
loss of generality, 151

manipulative materials, 33, 50, 64,
    68
mathematical induction, 134
measurement model, 89, 93, 103
memorization, 52
misconception, 7
missing factor, 90, 91, 103
modeling, 68, 117, 129
Monty Hall Dilemma, 215
multiplication table, 15

negative transfer, 241

organized list, 22, 112
outcome, 203, 215

paradox, 215
parity, 139, 197, 232
partition model, 82, 89
part-whole relationship, 81
Pascal, 18, 210
Pascal's triangle, 54, 117, 121
percent, 98
perception, 78
perimeter, 7, 58, 63, 229, 230
permutation, 21, 24, 46
physical model, 247
Piaget, 240
Pick, 168, 172
place value, 57, 58
play, 50
Plutarch, 63
prime number, 161
probability, 197, 205, 224
procedural performance, 57
productive thinking, 29, 177
proof, 151, 164, 242
proper fraction, 81
proportion, 97, 98, 244
proportional reasoning, 245
Pythagoras, 149
Pythagorean triples, 131, 230

randomness, 200
ratio, 96, 97, 198, 244
rational number, 99
recursive definition, 119, 122, 123, 173
reflective inquiry, 131, 144
relative frequency, 207, 219, 221
repeated addition, 14

rule of product, 19, 22, 27, 48, 49, 73, 134
rule of sum, 19, 26, 30, 54, 133, 134

sample space, 200, 202, 204, 214, 218
scalene triangle, 46
second-order symbolism, 35, 49, 51, 63, 64, 66, 73, 74, 98, 151, 199, 232, 238
simulated experiment, 208
spatial thinking, 179
special case, 11, 42, 151, 179
spreadsheet, 119, 123, 173, 180, 193, 208, 222, 225
subitizing, 57

tape diagram, 146, 147
text, 51, 55
The Geometer's Sketchpad, 75, 163, 165, 170, 172, 178, 188
Theon, 72
theoretical model, 247
theory of affordances, 73
three factorial, 20
trapezoidal number, 138, 143, 154, 157
trapezoidal representation, 138, 159, 160, 161, 202, 219
tree diagram, 20, 22, 73, 134, 202, 210, 216
trial and error, 69, 142
triangle inequality, 36, 38, 39, 40
triangular numbers, 107, 117, 125, 126, 136, 153, 248
triangulation, 188, 190

uncertainty, 51
unit, 46, 75, 76, 77, 83, 95, 169,
  229
unit fraction, 80, 85, 234

visualization, 72, 77, 173, 183,
  186, 193

Vygotsky, 33, 49, 50, 65

W^4S principle, 72, 73, 74, 91, 110,
  111, 248
whole, 88, 89
Wolfram Alpha, 107

Printed in the United States
By Bookmasters